Guide to Good Practice in the
Management of Time
in Complex Projects

Guide to Good Practice in the
Management of Time
in Complex Projects

The Chartered Institute of Building

CIOB

WILEY-BLACKWELL
A John Wiley & Sons, Ltd., Publication

This edition first published 2011
© 2011 The Chartered Institute of Building

Blackwell Publishing was acquired by John Wiley & Sons in February 2007. Blackwell's publishing program has been merged with Wiley's global Scientific, Technical and Medical business to form Wiley-Blackwell.

Registered office:
John Wiley & Sons Ltd, The Atrium, Southern Gate, Chichester, West Sussex, PO19 8SQ, UK

Editorial offices:
9600 Garsington Road, Oxford, OX4 2DQ, UK
The Atrium, Southern Gate, Chichester, West Sussex, PO19 8SQ, UK
2121 State Avenue, Ames, Iowa 50014-8300, USA

For details of our global editorial offices, for customer services and for information about how to apply for permission to reuse the copyright material in this book please see our website at www.wiley.com/wiley-blackwell.

The right of the author to be identified as the author of this work has been asserted in accordance with the UK Copyright, Designs and Patents Act 1988.

Library of Congress Cataloging-in-Publication Data
Guide to good practice in the management of time in complex projects / the Chartered Institute of Building.
    p. cm.
  Includes bibliographical references and index.
  ISBN 978-1-4443-3493-7 (alk. paper)
  1. Building–Superintendence. 2. Production scheduling. I. Chartered Institute of Building (Great Britain)
  TH438.4.G85 2011
  690.068′5–dc22
                        2010029193

A catalogue record for this book is available from the British Library.

This book is published in the following electronic formats: ePDF [9781444329605]; Wiley Online Library [9781444329599]; ePub [9781444329612]

Set in Gothic 720BT 10 on 13pt by Toppan Best-set Premedia Limited
Printed and bound in Malaysia by Vivar Printing Sdn Bhd

2 2013

Contents

Contents

Contents

Table of figures

Table of acronyms

ADM   Arrow diagramming method
ALAP   As late as possible
ASAP   As soon as possible
CBS   Cost breakdown system
CDM   The Construction (Design & Management) Regulations 2007
CPM   Critical path method
CTS   Count the squares
D&B   Design and build
DMS   Document management system
EPC   Engineer procure and construct
EVA   Earned value analysis
EVM   Earned value management
EXF   Expected finish
FF   Finish to finish
FNET   Finish no earlier than
FNLT   Finish no later than
FS   Finish to start
GMP   Guaranteed maximum price
ID   Identification data
MF   Mandatory finish
MS   Mandatory start
OBS   Organisational breakdown structure
PDF   Portable document format
PDM   Precedence diagramming method
SCL   Society of construction law
SF   Start to finish
SNET   Start no earlier than
SNLT   Start no later than
SS   Start to start
TF   Total float
WBS   Work breakdown structure
ZFF   Zero free float
ZTF   Zero total float

Time-Management Working-Group member and contributor details

Keith Pickavance LL.B. (Hons.), Dip.Arch., Dip.I.C.Arb., R.I.B.A., F.A.E., PPCIOB
Senior Vice-President, Hill International
Project leader, technical editor and contributor

Keith instigated the time-management initiative whilst in governance as president of the Chartered Institute of Building United Kingdom (2008–2009). He is a senior vice-president of Hill International and an architect by profession. His career spans more than thirty-five years in construction management, project planning, risk management, critical-path method of retrospective-delay analysis, and delay and disruption claims for building, civil engineering, oil and gas engineering, IT and shipbuilding contracts.

Alan Midgley B.Eng (Hons.), Civil, ICIOB
Associate Director Capita Symonds
Chairman of working group meetings and contributor

Alan graduated with a degree in civil engineering and has five years' postgraduate experience in temporary and permanent works design, site engineering and ten years' experience in planning. His experience is drawn from main contracting, project management and consultancy organisations. His main field of expertise is in the built environment and particularly in high-quality private residential schemes both in the UK and Singapore. Most recently Alan is heading up a team of planners delivering project controls and EVA for a £600 m mixed-use scheme from RIBA stage B to contract award.

Mark Russell BSc (Hons.). ICIOB
Technical Services Project Manager, NHBC
Working group contributor and project coordinator

Mark graduated with a 1st Class Honours degree in construction management from Sheffield Hallam University in 2005. To date he has had experience in the retail and residential construction sectors as a construction manager, building surveyor and building inspector. Mark acted as the project coordinator on behalf of the Chartered Institute of Building in the development of this 'Guide'.

David Tyerman MBA, LL.M
Planning Director, Spencer Engineering Group
Working group core member and contributor

David is planning director of Spencer Engineering Group, a civil engineering construction company, responsible for the effective development and control of all planning, scheduling and project controls. He also specialises in forensic analysis of project delay and disruption and has prepared numerous expert reports for clients spanning across a range of business sectors from construction to petrochemicals. He is passionate about the development of planning, scheduling and project controls as a core project-delivery discipline and of associated training, development and professional elevation.

Gildas André BSc (Hons.), MSc., MBA., MCIOB
Senior Manager, Ernst & Young
Working group core member and contributor

Gildas has been involved in all phases of the design and construction process and has extensive experience in development management, programme and project management, planning and scheduling and project finance. He currently provides management support to public and private clients in the UK and abroad on the planning and delivery of large capital investment programmes in the construction, real-estate, infrastructure and transportation sectors.

Paul Kidston PG Dip Building Management., MCIOB
Head of Planning, Vinci Construction Plc
Working group core member and contributor

Paul is experienced in production roles such as engineering manager, site manager, construction manager and planning manager. His experience spans a wide range of complex projects under differing contractual conditions, in a variety of sectors including high-profile public buildings, office developments and railways. Paul presents on earned-value analysis to various parties and client bodies such as the BAA and Boots and has given talks to professional bodies in the UK.

Robert Clark BSc (Hons.). MRICS., FCIOB
Private Consultant
Working group core member and contributor

Robert is a freelance consultant providing project management, IT and business financial consultancy services to UK public-sector and private-sector clients. His experience includes project and construction management, expert witness work, contract-implementation supervision, business planning and financial monitoring. His most recent appointment is that of special projects and change-management manager for a major public-sector client.

Tony Ciorra MAPM., CVA
Partner, Edge Consult (UK) LLP
Working group core member and contributor

Tony Ciorra's first ten years in construction were as a senior planner/project coordinator for contractors. For the past seventeen years he has been the principal consultant for a variety of commissions including both construction projects and non-construction business-change programmes. Tony gained his extensive experience on large construction projects in the commercial, health and retail sectors. His construction experience includes both design and construction expertise. He has also acted as an expert witness for major clients.

Trevor Drury PG Dip. Project Management, PG Dip. Law, MBA, FRICS, FCIOB, MCIArb
Managing Director, Estia Consulting
Working group core member and contributor

Trevor is a chartered quantity surveyor, project manager and dispute-resolution consultant with over twenty-seven years' construction-industry experience. He has worked for large civil engineering and building contractors, private practices, and contract and claims consultants. He currently specialises in project recovery providing interventions to successfully complete projects in delay and to manage the claims process in order to secure a successful financial settlement for the client. Trevor has also been appointed expert witness in relation to a number of legal proceedings both as quantum and project-management expert.

Earl Glenwright BS Civil Engineering. MBA
Member of the working group

Earl has a career spanning more than forty years in the construction industry. He is a certified planning and scheduling professional (PSP) and has been a member of the Project Management Institute College of Scheduling and the Association for the Advancement of Cost Engineering International Planning and Scheduling committees. His experience includes large and very large [super-mega] construction projects.

Patrick Weaver PMP., FAICD., FCIOB., MAIPM, MAPM
Managing Director Mosaic Project Services Pty, Ltd
Working group member

Patrick is a past president of the Chartered Institute of Building Australasia (2006– 2008) and has over thirty-five years' experience in the project management and construction industries. His career has been focused on planning and managing construction, engineering and infrastructure projects. In the last few years, his experience has also included developing PMOs and planning in a range of government ICT and business environments. Patrick has delivered a range of training courses and seminars around the world.

Other contributors:

Andrew Owenson

Dr Andrew Platten

Anthony Caletka

Ian Rollitt

John Banks

Octavian Dan

Peter Curtis

Peter J Green

Preface

The early days

I guess it all started in the autumn of 1974 when I was the architect in charge of a 54-bed nursing home project. I had written a letter to the nominated subcontractor for the mechanical installations criticising his performance and threatening to replace him unless things changed. In response I received a 17-page letter, criticising my spelling, grammar, syntax and understanding of the building contract, finishing with the words '*nemo dat quod non habit*', yours faithfully, legal director. That changed my life. Recognising that my education, as an architect, had not prepared me for plumbers writing to me in Latin and that, probably, I would have to suffer it for the next 30 years or so, I embarked upon a degree in law.

Construction management, earned-value analysis and computers

In 1982, by now in partnership with my brother, a quantity surveyor, the practice developed a facility in construction management as the procurement method of choice. The projects, which we designed and measured, were divided into separate trade packages and managed by us as the design team. In those days we scheduled the works for each package and their interfaces by bar charts, drawn by hand. The packages were usually small and rarely changed much, save in their timing and we used what is now referred to as earned-value management (EVM) to get a handle on how the progress in one package was likely to reflect upon the outturn cost and completion date of the whole, when several were being carried out in parallel.

We needed a computer to tie it all together but, in those days, we thought the PC (which then was an Apple 2b and which could not do much) would never catch on. So I mortgaged my pension fund and bought a Digital PDP 11-73 mini computer with three screens, two printers and a programmable database and spreadsheet. We thought it was pretty good in those days, notwithstanding that the spreadsheet could only calculate sequentially from top left to bottom right and, once the instruction to recalculate had been given, in relation to 40 or so linked spreadsheets, you could both make and drink a cup of tea before getting the result.

Critical-path analysis and a change of tack

By 1989 my brother had decided to become a contractor and I pursued a career as an architect, expert and sometimes an arbitrator. As an architect and expert in disputes, I had come to the conclusion that there was much that needed to be said for critical-path networks and their use in predicting consequences and, having woken up early one morning thinking about how an architect gives a fair and reasonable extension of time when there is both concurrent delay and float, I wrote a synopsis of a book which I felt needed to be written. My spare time, over the next seven years or so, was spent in reading everything I could find on CPM scheduling, risk management, proof of causation and so on which, to a great extent, became more focused after I joined an American consultancy in London in 1993.

In 1997, the first edition of *Delay and Disruption in Construction Contracts* was published. This was the first book on delay analysis published in the UK. At the time, the publisher and I both thought the book was about how architects could deal with extensions of time in the UK. But, within the first six months, it sold in 54 countries. Then we realised that time management was not parochial, it was an international problem, transcending cultural, jurisdictional and industrial issues, and affecting all industries in which a unique product is created, over a period of time, by a number of specialised designers, working under a contract that entitles the employer to change its mind about what it wants.

The 'delay analyst' became the name given to those who specialised in picking up the pieces after things had gone wrong. A whole new career structure then developed for architects (but not many), surveyors, civil engineers and project managers, who specialised in proving the cause of delays and their effect on completion, in troubled projects, in dispute, and many more books were to be written on the subject.

Training and education – the Masterclass series

By 1999, as a result of being employed doing little other than advising employers, contractors, subcontractors, architects and engineers on delayed projects, I had become increasingly concerned about the way the construction industry managed time – not just in the UK but internationally. It seemed that, on every project upon which I was then engaged, I saw the same incidence of poor scheduling, poor record keeping and poor project control. We almost always had to correct major deficiencies in what often were 'approved' schedules or 'programmes' as they tended to be called. Rarely had anyone, other than the contractor who prepared it, seen an electronic copy, and records, if kept at all, were often badly prepared and impossible to use, unless rekeyed into a database.

I decided to offer training to the industry in managing time and delay analysis and in 2000 set up a short training course called a 'Masterclass', the first of which, for about 32 people, was held at the PA Consulting conference centre in Bromley, Kent. It consisted of two days of lectures and case studies followed by a sample dispute to be analysed by syndicates in a workshop through the calculation of cause and effect, and presentation of case to an arbitrator or judge of the Construction and Technology Court. We paid the speakers well and charged a hefty price for the course but because of the costs of room and equipment hire (half a dozen computers, software and so on) we never made any profit on it. However, it was popular and we continued with it.

About a year later, I met John Douglas, the CEO of Englemere, who had heard of the Masterclass, and wondered if we could do it in conjunction with the CIOB. Undaunted by its unprofitability, John was of the same mind that profitability was not important so long as we didn't make a loss! We were concerned that few people could afford to take off from work three consecutive days and experience had shown that few of the delegates (more and more of whom seemed to be lawyers) had the computer skills necessary for the analytical work. We decided to shorten the Masterclass to two days, the first day in managing time proactively and the second in retrospective-delay analysis.

That is a successful formula which continues to this day where up to five Masterclasses, sometimes attracting as many as 90 people each, are now held in different parts of the world every year. So far, in partnership first with Pickavance Consulting and latterly with Hill International (who acquired my company in 2006), the CIOB has sponsored 16 Masterclasses in time management in various parts of England, three in each of Hong Kong and the UAE, two in each of Sydney and Dublin, and one in each of Brussels, Melbourne, Perth and Singapore.

The SCL Protocol and Change Management Supplements

At about the same time as the Masterclass series started, a drafting committee of the Society of Construction Law in London, of which I was privileged to be a member, embarked upon the drafting of a paper to promote good practice in the calculation of entitlement to an extension of time. In October 2002, what is now known as 'The SCL Protocol' was published.

The thesis of the Protocol was that, if the effect of events could be impacted upon a critical-path-method (CPM) network, which was up to date at the time, the effect could be calculated and measured contemporaneously, instead of being guessed. Moreover, this would be greatly to the advantage of everyone concerned with construction contracts because they could then manage their risks proactively instead of fighting about who would pay, after it was too late to put the problem right.

It was apparent that standard forms of contract did not encourage time management (typically, there were pages and pages of clauses dealing with cost but, if any at all, there would be just one clause dealing with time and that not linked to the extension-of-time provisions). Not only that, but, in some forms, effective time management was actually inhibited. Accordingly, in 2003, in conjunction with Fenwick Elliott, solicitors, I drafted a series of contract supplements for use with the 1998 series of JCT contracts to facilitate their use in the management of time.

Notwithstanding the obvious advantages, the industry did not take this message to heart. On the whole, contract-drafting bodies ignored both the SCL Protocol and the 'Change Management Supplements', as they were called, and unfortunately 'the Protocol', as it became known, was used more often as a stick with which to beat the opposition in disputes, rather than to manage time proactively and avoid disputes in the first place.

President of the CIOB

Disappointed by the absence of take-up of the recommendations of the Protocol and Change Management Supplements, and wondering what to do next about the continuing problems of poor planning, scheduling and project control, in the spring of 2006, I was invited to join John Douglas and Chris Blythe, the CEOs of Englemere and the CIOB, respectively, for lunch at a rather nice fish restaurant in Soho. They said they wanted me to make a bigger contribution to the CIOB and invited me to allow my name to go forward to council as a potential vice-president, and, if accepted, all things being equal, to become president of the CIOB in 2009.

They listened patiently while I argued that although, for the last 30 years or so, construction management had been the CIOB's cornerstone policy for improvement of the construction industry, time management (without which effective cost control was impossible) seemed largely to have been neglected. Ultimately, they agreed that if my nomination was approved, the CIOB would back my attempts to make improvements in the way the industry managed time.

CIOB research

Before embarking upon the process of finding a solution, it seemed to me that it was important that we had a good understanding of where the problems lay.

Following a period in which various research subjects and techniques were discussed, between December 2007 and January 2008, under my direction, the CIOB conducted a survey of the industry's knowledge and experience of different methods of project control and time management, under the heading of *Managing the Risk of Delayed Completion in the 21st Century.*

The thesis underpinning the research was that, despite the advice of the Protocol and availability of advanced computerised facilities, little had changed in the practice of time management since the development of the bar chart nearly 100 years ago. The essence of the research was thus to understand industry performance, the techniques used and the competence of those engaged in the process of time management. As far as we could ascertain, this was the first research of its kind.

The survey required the respondents to submit commercially sensitive information. Four-hundred companies were approached and 73 responses received, just under half of which were anonymous. The report was based upon data provided on nearly 2,000 projects over a three-year period.

The conclusions of the CIOB report

The survey showed that simple, repetitive, low-rise projects had a high chance of success within traditional, intuitive, time-management. However, the more complex the project, the less likely it was that an intuitive approach would be sufficient to achieve completion on time. Without a scientific approach to time management, complex buildings and engineering projects were likely to be substantially delayed in their completion.

The research revealed that the growth in training, education and skill levels of the industry in the use of time management techniques had not kept pace with the technology available. The overwhelming majority (95%) of the respondents thought that the standard of education and training in the management of time was unsatisfactory.

The Guide

From the research, it became apparent that time management in the construction industry in 2008 was in about the same state as cost management had been at the turn of the 20th century, about 100 years ago. There were no accepted standards to work to; no formal educational programme for those who set out to do it; no formal training for those doing it; and no accreditation, or qualifications to demonstrate competence.

However, it was apparent that, unless there were standards to which to work, there was little that could be done about education, training or accreditation.

Accordingly, the first stage was to write a guide to good practice. Even though there was much available in software training and several books and a recommended code of practice in retrospective-delay analysis, no one anywhere appeared to have attempted, hitherto, the writing of a guide as to how to manage time proactively and this looked as though it was likely to be a world first.

The working group

The first stage was to find a project coordinator and working group to put it together and eventually, in September 2008, Mark Russell was seconded from the NHBC as the group coordinator.

I wanted to get as many as possible, differently qualified enthusiasts in this field, from different backgrounds, from around the world, so that they could approach the problem from different directions. Advertisements for interest were placed in trade journals and a response page was placed on the CIOB website. I was invited to give the keynote address on the CIOB's research to the Construction SuperConference in London in May 2008 and to repeat the paper in Singapore in September for the Project Management Asia Conference. I took advantage of both to invite participation in the forthcoming Guide; it was as a result of the latter that Pat Weaver from Australia joined our group.

The first task of the working group was to consider what we mean by the words we use. This turned out to be an enormous task, which kept everyone busy for over five months. It produced a comparative table of all the definitions that various bodies used for the various terms, a document that would have produced a book on its own! Many signed up to the working group's website with an interest, and made valuable contributions from the outset. I persuaded an American chum of mine from the PMI College of scheduling, Earl Glenwright, to join the group and more signed up as interested parties but did not contribute. Those who stuck it out, formed the hard core of the working group and, from April 2009 onwards, much work was done in meetings around a table, either at the offices of EC Harris, or of Hill International, in London, together with the contributions from Pat in Australia and Earl in America.

The process we followed was that, first, someone who was interested in dealing with a particular aspect of the subject would write something and distribute it to the group via Mark. Mark would then field commentary on it from the rest of the group, sometimes via our discussion website, and an edited version would be tabled for group discussion.

By July 2009 we had started to draw together the various sections and it then fell to me to fill the gaps and to attempt to bring a common style to the various isolated contributions. As a result of this process, the core principles were developed together with the essential scheduling densities, strategies, quality assurance and the concept of the time-model. In September 2009 four days were set aside for the editing of the work and drawing the diagrams. David Thompson, a graphic artist, was introduced to the working group and he participated in several meetings, producing his wonderful free-hand sketches.

By the end of September, the Guide was ready for peer review and Mark and Sarah Naxton, the CIOB's Marketing Communications and Web Manager, worked tirelessly with Rob Clark to place the draft and questionnaire on the website, for downloading and commentary.

January 2010 came around and it was time to look at the results of the peer review. Over 200 copies of the draft had been downloaded for comment and many extensive and informed commentaries had to be reviewed and the Guide edited to account for the suggested improvements. A further series of meetings through February 2010 produced further drafts and by March 2010, the last draft was deemed fit to be sent for print.

It has been a long, but extraordinarily satisfying journey. The working group, enthusiastically supported by Earl, in America, and Pat, in Australia, have had many passionate and fascinating debates. Some have gone on for many meetings and have resulted in several drafts and re-drafts before a consensus could be achieved. However, we all recognise that this is just the beginning. Although it is the best we can do now, I have no doubt that before long it will need to be revised and updated as a result of more experience and better minds turning to this subject.

The next stages of the development of time management will be in setting up an educational and training structure and providing some form of accreditation for those who achieve an appropriate level of competence in time management.

Not only will the construction industry benefit from this but, because they tend to suffer from the same difficulties, we hope that it will also be of some benefit to the industries of civil engineering, water, gas and oil, IT and shipbuilding, amongst others.

Keith Pickavance PPCIOB
Ascot, 2010

The Time Management Working Group, September 2009
Back row: Paul Kidston, David Tyerman, Gildas André, Mark Russell
Front row: Rob Clark, Trevor Drury, Alan Midgley, Keith Pickavance, Tony Ciorra

Acknowledgements

A special thanks to the following:

EC Harris LLP and Hill International for use of their London offices

Saleem Akram: Director of Construction Innovation and Development, The Chartered Institute of Building

Toby Hunt: Senior Vice-President, Hill International, for his assistance with process management, external relations and coordination of research

Sarah Naxton, Marketing Communications and Web Manager, Policy and External Relations, The Chartered Institute of Building

David Thompson: Illustrator

And all those who responded to the consultation document, but, in particular, thanks to the following for their extensive and detailed review:

David Stockdale
John Hayward
Murray B Woolf
Raf Dua

1 Preamble

1.1 Core principles

1.1.1   The Guide is a practical treatise on the processes to be followed and standards to be achieved in the effective management of time. It is not based upon any contractual regime, or procurement process and, subject to amendment of existing forms of contract to remove inconsistencies, can be used in any jurisdiction, under any form of contract, with any type of project.

1.1.2   Without effective time management there can be no effective resource management, cost management, nor allocation of liability for slippage, its recovery, or accountability.

1.1.3   In order to achieve effective time management there must be:

- ■   a competent appraisal of the risks which are likely to severely disrupt and delay the progress of the work;

- ■   a design which permits the work sequences that are likely to be severely disrupted and delayed by foreseeable events to be separated into parallel, rather than sequential paths;

- ■   a 'time-model' for the project against which progress, or lack of it, can be measured;

- ■   a practically achievable strategy for dealing with intervening events during the design, procurement and construction processes.

1.1.4   The word 'programme', often used in the past to describe a printed paper copy of a listed process and dates on which the proposed activities might be carried out, is not used in connection with the management of time in complex projects.

1.1.5   The word 'schedule' is used to describe the computerised calculated activity dates and logic; the process is to be referred to as scheduling and the occupation that of the scheduler. It is a process manifest in an editable computer file.

1.1.6   Planning and scheduling are separate disciplines. Project planning is largely an experience-based art, a group process requiring contribution from all affected parties for its success. On the other hand, scheduling is the science of using mathematical calculations and logic to predict when and where work is to be carried out in an efficient and time-effective sequence.

1.1.7   Planning must precede scheduling. They cannot be carried out in parallel, nor can scheduling precede planning.

Guide to Good Practice in the Management of Time in Complex Projects, by The Chartered Institute of Building
© 2011 The Chartered Institute of Building

1.1.8   Schedule preparation must be a quality-assured process against a standard which will ensure the integrity of the schedule, so that it can function as a time-model.

1.1.9   The schedule (and any revisions and updates) is to be independently audited for integrity and technical competence.

1.1.10   Time management starts on the drawing board with the conceptual design. If the design is not time-effective, no procurement strategy will rescue it.

1.1.11   Time management of complex projects necessarily encompasses the management of design, manufacture, procurement, subcontract and separate contractor work packages, information flow, quality control, safety management and the achievement of multiple key dates, sectional completion dates and multiple projects.

1.1.12   A time-risk appraisal is to be carried out at inception and constantly updated throughout the life of the project.

1.1.13   Time contingencies for the employer's, the design team's and the contractor's risks must be a part of the strategy for effective time control.

1.1.14   The Guide differentiates between the Development Schedule, prepared before a contractor is appointed, and the Working Schedule used in connection with construction.

1.1.15   The Development Schedule cannot be prepared in one process at a single density, or degree of detail at inception. It must be prepared in varying densities consistent with the information available from time to time, and reviewed and revised at regular intervals, as better and more certain information becomes available.

1.1.16   The Working Schedule must follow from the Development Schedule and must also be prepared in varying densities consistent with the information available from time to time. It must also be reviewed and revised at regular intervals as better and more certain information becomes available.

1.1.17   Consultants', specialist contractors' and subcontractors' schedules are to be prepared in the same software as the Development and Working Schedules and integrated into them.

1.1.18   Progress monitoring techniques, which are rooted in comparison of data against a static baseline, have limited value in competent time management in complex projects (in which the work content, resources and sequence necessarily change from time to time).

1.1.19   The work to be carried out in the short term must be scheduled according to the resources to be provided and the productivity quotients for the various work types to be carried out. The absence of a high-density, short-term part of the schedule, or a short-term part calculated other than by reference to resources, is not permitted under this Guide.

1.1.20   Because progress data will be entered only against a fully resourced schedule, the as-built record will provide data standards and productivity feedback for future benchmarking, which will improve predictability and hence reliability of short-term scheduling.

1.1.21   Progress records are to be kept on a database which will provide instantaneous access and retrievability of as-built data for the purpose of checking the reliability of productivity assessments in varying repetitive work cycles.

1.1.22   Quality control and information flow should be managed via the same database as that used for the maintenance of progress records.

1.1.23   The effective management of time necessarily includes the management of the consequences of delaying events.

1.1.24   Intervening events are to be impacted at the time of their initiation, along the lines recommended by the SCL Protocol[1]. The likely consequences of intervening events are to be calculated.

1.1.25   There is no guidance for the approximation of a 'fair and reasonable extension of time', nor of likely delay-related cost claims.

1.1.26   The time management strategy is to be set down in writing in a regularly updated method statement, which is to deal with, amongst other things, the stated strategy and assumptions adopted for:

- project planning

- risk management

- schedule preparation

- schedule review and revision

- progress update

- record keeping

- quality control

- communications

1.2   Mission statement

1.2.1   The primary purpose of this Guide is to set down the standards necessary to facilitate the effective and competent management of time in construction projects.

1.2.2   The Guide defines the standards by which project schedules will be prepared, quality controlled, updated, reviewed and revised in practice.

1.2.3   The Guide describes the standards of performance which should reasonably be required of a project scheduler. It will also form the basis for the education of project schedulers.

1.2.4   Without compromising its primary purpose, the Guide will be developed as a scheduling reference document capable of wide application.

1.3   Genesis of the Guide

1.3.1   The continuous pursuit of excellence in the management of construction is the key to greater effective collaboration, the continued satisfaction of the industry's client requirements and the sustained delivery of successful projects in the 21st century.

1.3.2   With a view to examining the state of the industry in time management, between December 2007 and January 2008 the CIOB conducted a survey of the industry's knowledge and experience of different methods of project control and

---

[1] The Society of Construction Law, Delay and Disruption Protocol (2002).

time management, record keeping, monitoring and training.[2] The results indicated a wide disparity between the experience of the respondents and good practice in time management.

1.3.3   In the light of the results of that research and with a view to reducing the incidence of delayed projects, the CIOB has initiated this Guide as a part of the initiative in encouraging excellence in the management of construction, increasing awareness of the importance of project planning and scheduling in the industry as a whole and, in particular, with regard to the management of time in complex projects.

1.4   Purpose of the Guide

1.4.1   The growth in training, education and skill levels of the industry in the use of time management techniques has not kept pace with the technology available. There is, however, a trend towards developing more complex projects with contracts which, if not executed efficiently, with good-quality time management and project controls, are increasingly punishing.

1.4.2   It is apparent that, since the 1980s, the construction industry has experienced:

■   Design and Build, Guaranteed Maximum Price, and Engineer, Procure and Construct Contracts and other devices, which require the contractor to take more risk than in traditional forms of contracting;

■   specially incorporated companies as employer for a particular project which will have limited access to additional funds and are intended to be liquidated once their purpose is fulfilled;

■   more technologically complex solutions in shorter timescales and within tighter financial constraints.

1.4.3   The purpose of this Guide is thus to set down the strategy and the standards necessary in order to facilitate the effective and competent management of time in complex projects.

1.4.4   This is not a guide to project risk, value, or other management specialities.

1.4.5   Time-modelling with the use of computers to develop a framework by which the consequences of change and other intervening occurrences can be managed technically and objectively has been available since the early 1960s. However, it is only in the last few years of the 20th century that the necessary computing power and software have become commonly available to facilitate the objective measurements of project deliverables, except in the most unusual circumstances.

1.4.6   Developments in hardware, software and communications' services in the last decade of the 20th century have rendered it virtually impossible in the 21st century to conduct any business efficiently without the use of computers and electronic services.

1.4.7   At the time of writing this Guide, it is apparent that the construction industry uses those resources intensively in design, in manufacture, in procurement, in assembly, in finance and in virtually every field other than the management of time. It is apparent, from the CIOB's research, that time management is generally

---

[2] See *Managing the Risk of Delayed Completion in the 21st Century*, 2008, Chartered Institute of Building (downloadable from http://www.ciob.org.uk/filegrab/TM_report_full_web.pdf?ref=880 [accessed 14 August 2010]).

pursued intuitively and schedules, if used at all, are used only as a target against which failure to succeed can be reported.

1.4.8   Whilst it is apparent that simple projects can be managed intuitively by experienced construction managers, it is also apparent that the management of complex projects cannot. Attempts to manage time on complex projects by intuition alone will result in failure. In complex projects there are simply too many consequential possibilities for time to be managed by intuition alone. A more scientific approach is required to assess the consequences of express and implied changes and the effect of other intervening events on the multiplicity of activities in a changing time frame.

1.4.9   However, just because the Guide focuses on the requirements of complex projects, this does not mean that what is recommended here cannot be adopted for simple projects if that is what the employer and/or contractor or consultants wish in particular circumstances. On the other hand, it does mean that managing time by intuition alone is simply not good enough for complex projects.

1.5   Applicability of the Guide

1.5.1   Complex projects can be defined both inclusively and exclusively. Much will depend upon the perceptions of those involved as to whether, for the purpose of time management, a particular project is simple or complex. On the other hand, experience and the results of the CIOB's research indicate that the following classifications are likely to prove helpful.

1.5.2   Simple projects

1.5.2.1   Simple projects comprise those in which construction has **all** the following characteristics:

- design work is completed before construction starts;
- work comprises a single building (or repetition of identical buildings);
- construction is lower than 5-storey height;
- without below-ground accommodation;
- carried out to a single completion date;
- without phased possessions, or access dates;
- with services not exceeding single-voltage power, lighting, telephone, hot and cold water, and heating;
- with a construction period shorter than 9 months;
- with a single contractor; and
- with fewer than 10 subcontracts.

1.5.3   Complex projects

1.5.3.1   Complex projects comprise those in which construction has **any** one or more of the following characteristics:

- design work is to be completed during construction;
- work comprises more than one building;
- construction is higher than 5-storey height;
- contains below-ground accommodation;
- to be completed by multiple key dates and/or sectional completion dates;

- with multiple possessions, or access dates;

- with short possessions;

- work contains services exceeding single-voltage power, lighting, telephone, hot and cold water, and heating;

- construction work is accompanied by work of civil engineering character; or

- the construction period is longer than 12 months;

- construction is to be carried out by multiple contractors; or

- by more than 20 subcontractors.

1.6 Planning and scheduling

1.6.1   Project planning and scheduling, although they are allied disciplines, are not one and the same.

1.6.2   In principle, project planning is a team operation, involving the construction management team, cost control team, design team and project planner in creating the project development strategy. There are fundamental aspects of planning which require a conceptual approach similar to designing. It requires experience, vocabulary, communication and imagination and, at its highest level, provides the formula for the logistic strategy for the project construction.

1.6.3   Project planning involves decisions concerning:

- the overall strategy of how the work process is to be broken down for control;

- how the control is to be managed;

- what methods are to be used for design, procurement and construction;

- the strategy for subcontracting and procurement;

- the interface between the various participants;

- the zones of operation and their interface;

- maximising efficiency of the project strategy with respect to cost and time;

- risk and opportunity management.

1.6.4   On the other hand, scheduling is a mixture of art and science, involving the interpretation of the results of project planning to ascertain, amongst other things, the start and finish dates of activities and their sequence. Scheduling is usually performed by the use of software which facilitates the fast and efficient manipulation of the project planning data for the purposes of time and risk management. In effect, the schedule is the construction manager's time-allocation tool, the employer's risk-management tool and the contract administrator's calculator.

1.6.5   Project scheduling is the art and science of putting the decisions made at the project planning stage into a database:

- to enable the scheduler to allocate contract calendar periods to the various sections of the work in a logical sequence;

- to allocate contingency periods;

- to make decisions as to preferred sequences;

■ to calculate float in relation to resources available;

■ to present the strategy in a form acceptable to the contractor, employer and the contract administrator as a process-management tool.

1.6.6   In the process of converting the plan into a schedule (within a framework, which will react dynamically to change, so as to facilitate the management of time throughout the life of the project), the scheduler should determine:

■ the duration of the activities;

■ the party who will perform the activities;

■ the resources to be applied to the activities; and

■ the method of sequencing of one or more activities in relation to other activities.

1.6.7   It is not good practice to plan the work whilst attempting to schedule it. In the same way that it is possible to start designing a building at the same time as preparing the working drawings and other production information, it is equally possible to perform the project-planning operation whilst scheduling. However, in neither case is such an approach likely to produce, on the one hand, a satisfactory design and consistent production information nor, on the other, a satisfactory project-planning solution and effective schedule.

1.6.8   Accordingly, the Guide recommends that the project-planning function is performed first and the scheduling operation carried out in accordance with the established strategic project plan and planning method statement.

1.6.9   In essence, the prior planning procedure should encompass:

■ familiarisation

■ outline plan

■ strategic plan

■ detailed plan and planning method statement

1.6.10   Only when that has been completed should the project-scheduling process commence.

1.6.11   The importance of the project schedule to time management cannot be overemphasised. Without a dynamic time-model which will react dynamically to change, it is not possible, except intuitively, to forecast when work is to be carried out, nor is it possible to assess its criticality, nor the impact on successor activities, nor resources.

1.7   The project scheduler

1.7.1   The job of the project scheduler is to devise and maintain the process plan from overview to micro-level and to manage practical and effective time control from commencement to completion of construction projects.

1.7.2   In order to control the probity of updated schedules, the scheduler must thus be able to advise on and manage the making and retrieval of progress records and, in order to keep the project and construction-management team informed, the project scheduler must be able to prepare as-built schedules of work carried out, progress schedules and progress-related data, for the purpose of progress reporting during the course of the works.

1.7.3   Apart from writing the schedule at inception, the project scheduler will thus be engaged in writing, editing, reviewing, revising and updating the schedule. Reviewing and revising schedules in the light of better information bring the need also to be able to advise on and manage the writing, revision and editing of project-planning method statements.

1.7.4   When change is imposed, the project scheduler must also be able to identify contemporaneously the effect of delaying and disrupting causal events on the planned sequence and to advise the project planner and other members of the project-management team on the likely effect of possible recovery strategies.

1.8   Project control

1.8.1   Project control is the science of identifying, from time to time, what, in the light of current status and information, the completion of a sequence, key date, sectional completion date or completion date is likely to be and then, if that is not what is required, in the light of the information then available, amending the strategy and schedule for the future conduct of the work so as to plan to achieve what is required.

1.8.2   Accordingly, the schedule is to be used for identifying, from time to time, the following intentions:

- periods of activity and sequence of the work and the interface with any other contracts incidental to the work;

- dates and logic by which the information described in the information-release schedule, information-request schedule, or any other request for information, is to be supplied in relation to the activity dependent upon such information;

- dates and logic by which plant, materials or goods are to be supplied, or work to be carried out by the employer, or those engaged or employed by them in relation to the activity dependent upon them;

- any time contingency required by the contractor, any subcontractor and/or supplier in relation to any activity, sequence of activities, or key dates, or the contract requirements for any sectional completion dates and the completion date;

- any time contingency required by the employer, or any directly employed contractor, or consultant, in relation to any activity, sequence of activities, key dates, the contract requirements for any sectional completion dates and the completion date;

- free float and total float that are available to be used by the contractor and/or the employer for managing the re-sequencing of the work or redeployment of resources from time to time;

- the degree of progress actually achieved, on all activities, from time to time;

- the likely and actual effect of any delay to progress on the completion of any sequence, key dates, the contract requirements for any sectional completion dates and completion date, if any, caused by a change, or other intervening event;

- the likely effect of any proposed acceleration, or recovery measures on any such sequence, key dates, the contract requirements for any sectional completion dates and completion date.

2 Strategy

2.1 Planning

2.1.1   An effective time-management strategy will recognise that time expires at a regular and consistent rate, from inception to completion, whether it is used effectively, or not used at all. Accordingly, an effective planning strategy will demonstrate the most effective use of available time, in all circumstances.

2.1.2   Major projects are necessarily conceived a long time before they are designed; designed a long time before they are constructed; and put into use a long time after construction starts.

2.1.3   The larger and more complex the project, then the greater will be the time between inception and completion, and the more likely it is that there will be changes and other intervening events to be taken into consideration in the future.

2.1.4   A planning strategy which facilitates the effective management of changing subject matter, work content, sequences and resources and other intervening events is thus a prerequisite of the effective management of time on major projects.

2.1.5   The most effective time-management strategy starts in the design stages of a project. In the same way that, to some extent, it is possible on all projects to identify a cost-effective way of achieving the same quality, so projects can be designed to be time-effective without compromising out-turn cost, or quality.

2.1.6   If time-management considerations have not been entertained during the design stages of a project, then the opportunities for effective management of change and other impeding events may be limited during the construction stages.

2.1.7   In order to achieve the most effective time-management strategy, the employer, design team, contractor and subcontractors are to have the opportunity to contribute to the effective planning of the part, or parts, of the project with which they are concerned.

2.1.8   Where it is foreseeable that the occurrence of a predictable risk will severely delay a sequence, a time-effective planning strategy will take into account the likelihood of that risk occurring.

2.1.9   As part of the evolution of the planning method statement, the planning strategy is to be reviewed and articulated before scheduling commences.

2.1.10 The planning strategy is to be reviewed and revised from time to time, in the light of the occurrence of events not previously taken into account.

Guide to Good Practice in the Management of Time in Complex Projects, by The Chartered Institute of Building
© 2011 The Chartered Institute of Building

2.2   Schedule preparation

2.2.1   The overarching purpose of a schedule is to indicate when work is to be performed in the future. The strategy for effective schedule preparation must be to provide a prediction of what is the intended timing and sequence of work yet to be carried out. In other words, it must show how the work is planned to happen and be a predictive, practical model for the future conduct of the work.

2.2.2   The preparation of a competent and effective schedule must be based upon a sound, time-effective, planning strategy and must not proceed until the planning method statement has been developed.

2.2.3   The schedule objectives, structure and layout should be designed before scheduling commences. The content of the schedule will change during the development of the design and construction of the project. Accordingly, the schedule is to be designed to accommodate change in subject matter, content, method and source of data without compromising transparency between the schedule and other time-related information.

2.2.4   Because the content and timing of the work are unlikely to remain unchanged for more than a brief period, the schedule must be designed to accommodate changes. The schedule must predict the consequences of any change.

2.2.5   There must be an effective audit trail between the data on the schedule and the planning method statement, identifying why it is planned that way.

2.3   Schedule review

2.3.1   The choices available will vary throughout the course of the project, as will the perception of importance of the criteria driving the decision-making process. For example, notwithstanding that, at tender stage, there may be a method provided for aggregating cost, or for identifying time allowances, in complex projects it is unlikely that full details of all the work actually to be performed will be available at that time and much may remain to be designed, or decided upon during the construction process.

2.3.2   Accordingly, whatever choices are made and/or decisions taken at any moment, as to their contribution to effective time management, they can never be any better than the information upon which they are based. It follows that if, when decisions are made, the information upon which they are based is sketchy and ill-defined, notwithstanding that the choices may be precisely described and the decisions recorded, they may turn out to be poor in the light of later and better information.

2.3.3   The strategy for schedule review must take account of the development of the schedule as better information becomes available and, as the project proceeds, the increasing density of the schedule as it develops from initiation through the work on site to commissioning the completed project.

2.3.4   Irrespective of whether approval of revisions is required contractually, the scheduler is to ensure that no schedule revision is to constitute, or lead to:

■   an illegal operation; or

■   a breach of contract; or

■   a hazard to health and/or safety in the process of construction; or

■   a hazard to the safety and/or stability of the permanent work; or

■ a method or sequence which is not conducive to effective time control; or

■ misrepresentation, or otherwise unethical conduct.

2.3.5 Any positive control of revision to the schedule should encourage safety, contract and legal compliance and effective time control.

2.3.6 In so far as a submittal of a revision is required to be made for any reason, it is to be submitted as soon as the proposals for revision are defined.

2.3.7 A submittal for approval of a proposed revision is to be considered promptly and either approved (in which case the revised schedule becomes the working schedule for the future of the project) or rejected. The criteria for rejection are limited to those set out above, against which full details of the reasons for any such rejection should be given.

2.4 Progress update

2.4.1 The purpose of a progress update is to demonstrate the effect of progress achieved on the timing and sequence of work still to be carried out. Because the sequence and timing of future work necessarily are dependent upon the degree of progress made and capable of being made with the available resources, it is essential, for effective time management, that the schedule is updated at regular intervals.

2.4.2 The appropriate interval of schedule update should be consistent with, and should not exceed, reporting periods.

2.4.3 Each schedule update is to identify the status date or 'time now' at which the progress of the work is recorded. All work started is to show a start date to the left of the data date and all work to be performed in the future must be to the right of that. No other formula is acceptable for demonstrating effects on progress of timing and sequence.

2.4.4 Where any activity has started or finished out of sequence, the logic of the schedule and planning method statement must be amended to demonstrate the logic which was actually followed and the reason for the change.

2.5 Change management

2.5.1 Change is anything which is not planned and it can have both positive and negative effects on the schedule.

2.5.2 Change-risk management requires that the risks likely to affect the progress of the work from inception to completion are identified as early as possible, together with a meaningful strategy for dealing with the possibility of their occurrence.

2.5.3 Thus, an effective delay-risk-management strategy, in relation to any given planning strategy, should be a review of the answers to the following questions:

■ what can go wrong?

■ what is the likelihood of that happening?

■ when is it likely to happen?

■ is it likely to have a consequential effect on a key date, or completion?

■ can the planning strategy be reconceived to avoid the possibility of the risk occurring?

■ can the planning strategy be reconceived to avoid the possibility of the risk, if it occurs, having a knock-on effect on completion?

■ does that require a time contingency, or the adjustment of a contingency?

■ if not, how is the likelihood of the occurrence of the risk to be catered for?

2.5.4 A project-risk register will fulfil the useful purpose of identifying those occurrences which reasonably can be foreseen as being likely to have an adverse effect on progress, and what can be done to minimise the occurrence of those risks having an adverse effect on completion.

2.5.5 Once the master risk register has been set up, it must be reviewed regularly against emerging risks and, where necessary, revised by reassessment of priorities and foreseeable risks through design, tender and construction, including services' appointments and employer procurement through to testing, commissioning and occupation.

2.5.6 Whilst there cannot be a time-effective planning strategy unless the risks which are likely to affect the future conduct of the work are allowed for, without a planning strategy for the future conduct of the work, it is impossible properly to appraise the risks likely to affect it. It thus follows that the development of an effective delay-risk-management strategy is inseparable from a time-effective planning strategy.

2.5.7 In relation to those risks which cannot be avoided entirely, the purpose of a change-risk-management strategy must be to facilitate the recovery of lost time economically and effectively. This is usually achieved when carried out as a result of strategically identifying time contingency buffers against the foreseeable risk of delay to progress in critical zones.

2.5.8 During the early stages of every project there are many alternative sequences that can be pursued. However, those options reduce as the project proceeds and it is unusual to find a project with many opportunities for re-sequencing in the later stages.

2.5.9 A typical profile of the risk of delay to the start or finish of discrete activities and the opportunity and cost escalation of dealing with their consequences can be illustrated as Figure 1, below:

2.5.10 However, it is obvious that, no matter how good the management, it is not possible to 'manage-out' the likelihood of occurrence of all risks and this should not be seen as a substitute for a change-risk management strategy, which can account for those risks which have a reasonable probability of maturing into intervening events. Typically, but dependent upon the contract and the project, those risks for which allowances should be considered are listed in Appendix 1.

2.5.11 A shift in timing is almost certain to occur as a result of the maturation of one or more foreseeable risks into an intervening event at some time during the life of the project. On a complex project it is indefensible to proceed on the basis that an intervening event will not occur or, if it does, its effect on progress (and its consequences) can be dealt with intuitively at the time.

2.5.12 If an intervening event takes up time and/or resources which do not form part of the original agreed contractual obligations, records must identify that intervening event and the work that flows from it.

2.5.13 The work content, timing and sequence involved in an intervening event are to be estimated and added to the schedule at the earliest opportunity and the schedule is to be updated with factual information as it becomes available.

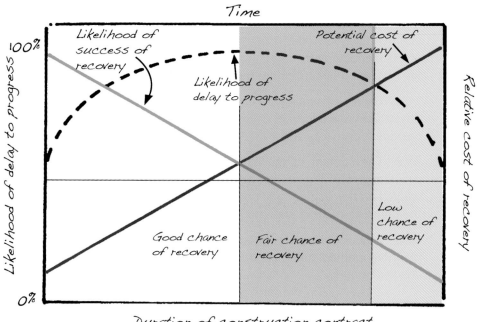

Figure 1 Profile of risk of delay to progress.

2.6 Planning method statement

2.6.1 Consultation and effective communication are prerequisites of a meaningful and effective schedule and planning method statement, defining the strategy of how the project will be executed and managed. At inception the planner will draft an appropriate planning method statement with the input and cooperation of others. Those who may participate will depend upon the nature of the project, but might reasonably include:

- the employer
- the design team
- the risk manager
- the project manager
- the construction manager
- the project scheduler
- the health and safety planning manager

2.6.2 As the density of the planning method statement increases, so the input of others will be required, for example:

- other directly employed contractors
- utilities, statutory undertakers and third-party projects
- specialist designing subcontractors
- trade-package contractors
- domestic subcontractors
- specialist suppliers

2.6.3   The purpose of the planning method statement is to facilitate the understanding and cooperation of the participants. It should make clear what constraints have been identified, what assumptions have been made in the process of risk management, planning, scheduling, review and update of the schedule and the reasoning underpinning those constraints and choices.

2.6.4   Because it will have a life independent of those who, from time to time, may be required to work upon it, it is important that the planning method statement is designed for use by those independent of the project.

2.6.5   The content of the planning method statement will change during the development of the project and must be designed to accommodate change in subject matter, content and source, without compromising transparency between the planning method statement and other time-related information.

2.6.6   Any amendments made to the underlying assumptions contained in the planning method statement will also need to be carefully recorded in a clear and concise manner.

2.7   Record keeping

2.7.1   The purpose of records is to provide an audit trail of work carried out.

2.7.2   Effective and meaningful records are important because they are:

■   the factual evidence against which the schedule is updated;

■   the proof of the productivity quotients used to plan the work;

■   the factual basis for the productivity assumptions for the future of the project;

■   the basis for demonstrating cause and effect.

2.7.3   Effective records are those which can easily be understood, accessed, sorted, filtered and reported upon. It is thus essential that they be kept electronically, as database records.

2.7.4   There must be an effective audit trail between the data on the schedule and the records of work actually performed by the resources actually employed.

2.7.5   Except in the most unusual circumstances, records are to be made and maintained at regular intervals. Depending upon the project, work type and types of record, that may mean monthly, weekly, daily, or hourly.

2.7.6   A process is to be identified by which inconsistent data can be avoided and consistent data can be agreed.

2.8   Time-management quality control

2.8.1   The purpose of the quality-control process is to ensure that the recommended practices in this Guide are followed and that any departure from the recommended practices is adequately documented with:

■   the reason for such departure;

■   the acceptance by the contracting parties of the need for such departure;

■   recognition of the likely consequences of the departure.

2.8.2   There is to be an effective quality-control process, independent of any contractual third-party review which, contractually, may be required. The quality-control process is to be specifically designed to review the schedule, planning method statement, schedule revision, schedule update and record-keeping processes.

2.9 Communications

2.9.1 Effective communication requires that all parties are in possession of a common data set, such that identified emergent risks may be managed and alternatives for future performance considered in the most timely and effective manner. This requires that all parties are in possession of:

■ project scope and objectives;

■ the planning method statement explaining the strategies for the project;

■ the schedule;

■ contemporaneous progress records; and

■ the risk register.

2.9.2 The employer, contractor, consultants and others responsible for any work performed under the contract should have equal access to the same information regarding sequencing and timing of the work to be performed.

2.9.3 All time-related information and data are to be produced and made available electronically.

2.9.4 There must be a common denominator linking information of like type between different databases and documents (e.g. the activity ID used by the project scheduling software).

2 Strategy

3 Developing the time-model

3.1 Introduction

3.1.1   In any one project, there will be many parties with a legitimate interest in ensuring that the timing of the work is managed effectively. Typically, these will include:

- those financing the project
- the employer
- the contractors
- the subcontractors
- the suppliers
- the design consultants
- the project manager
- the contract administrator

3.1.2   The purpose of the time-model is to indicate when in the future and in what sequence the planned work is to be performed, so that the intended work and the consequences of any changes, or departures from that intention can be predicted, communicated and managed efficiently.

3.1.3   Because, at any one time, the time-model can only be as accurate a prediction of the future as current knowledge will allow, it must be conceived as a model which can be improved upon as information becomes available or circumstances change.

3.1.4   In order to facilitate efficient time management, the time-model should be constructed so as to differentiate between work that can be predicted:

- in outline in the long term;
- in detail but with some information missing in the medium term; and
- accurately as to the content sequence and resources to be employed on work which will be carried out in the short term.

3.2 Developing the schedule

3.2.1   Whilst every project will have its own determining characteristics, the bulleted considerations, below, identify those general matters which, ordinarily, should be considered in designing the schedule:

Guide to Good Practice in the Management of Time in Complex Projects, by The Chartered Institute of Building
© 2011 The Chartered Institute of Building

■ Time for completion	■ Licences and permissions
■ Sectional and key completion dates	■ Provisional and prime cost sums
■ Unspecified milestones	■ Specifications
■ Access, egress and possessions	■ Bills of quantities
■ Information-release dates	■ Local regulations
■ Submittals and approvals	■ Environmental conditions
■ Procurement strategy	■ Health and safety
■ Procurement schedule	■ Noise restrictions
■ Materials' delivery and storage	■ Labour and plant resources
■ Temporary works	■ Logistics
■ Temporary traffic arrangements	■ Construction philosophy
■ Working hours and holidays	■ Method of construction
■ Design responsibility	■ Sequence of construction
■ Complexity of design	■ Schedule requirements
■ Adjoining owners	■ Updating requirements
■ Risk allocation	■ Notice requirements
■ Subcontractors and suppliers	■ Reporting requirements
■ Separate contractors	■ End-user requirements
■ Employer's contractors	■ Testing and commissioning
■ Employer's goods and materials	■ Furniture and fittings
■ Nominated subcontractors	■ Phased occupation
■ Utilities and statutory undertakings	■ Occupation and handover
■ Third-party issues	■ Partial possessions

3.3   Schedule types

3.3.1   Theoretically, there is no limit to the number of different types of schedule that can be produced and, historically, there has been an understandable tendency to devise a new ad hoc type of schedule for every different purpose. However, this is an approach to be avoided.

3.3.2   Typically, there should be no more than five schedules. The element which distinguishes these schedules, one from the other, is the parties who participate in their creation. Every purpose to which they are to be put should be satisfied by organisation and filtering of the schedule type relevant at the time. The schedule types are:

■   development

■   tender

■   working

■   occupational commissioning

■   as-built

3.3.3   **Development schedule**

■   This is the schedule worked to by the employer and its consultants before any contractor is involved. Its focus should be the work needed to be undertaken by the employer leading into the contract and in support of the contractor, including the design work and obtaining the approvals necessary for the contract to proceed. The work of the contractor should be included at an appropriate level of detail and the schedule should follow the same rules as to density as any other schedule. (For example, at inception the anticipated construction period for the works on site may be represented, at low density, by a single bar; however, by the time the scheme design has evolved, much of the anticipated construction process will be identified in medium density.)

3.3.4 Tender schedule

■ This is the first schedule produced by the contractor. It will incorporate any information required by the design team (transferred from the development schedule) and illustrate, at a mixture of low and medium density, what the contractor intends to achieve and when.

3.3.5 Working schedule

■ This is a refined version of the tender schedule and will be the schedule used for planning and progressing work on site from commencement to completion. It must be completed in high density for the first three months of the project, before work on site commences.

3.3.6 Occupational commissioning schedule

■ This is the employer's schedule as to how it will occupy and use the building. It will contain details of possessions, furnishing, commissioning and testing, access and training. The schedule will usually be prepared by the employer or its consultants from the working schedule and reviewed and revised in the light of changes made during the course of construction.

3.3.7 As-built schedule

■ This is the final schedule to be completed. It will evolve through the course of the project as activities are started, progressed and completed as the work is executed. The last update should complete the as-built schedule as a high-density record of the sequence in which the works were actually constructed, the resources actually used and the productivity actually achieved.

3.3.8 In the same way that the contractor's tender schedule is expected to be derived from the development schedule, so the subcontractor's tender schedule will also be derived from the development schedule, or the contractor's working schedule depending upon the timing of the tender. After the subcontract has been agreed, the subcontractor's tender schedule should then be matured and incorporated into the working schedule.

3.4 Scheduling techniques

3.4.1 The range of scheduling techniques available and their application for use in an effective time-model will largely depend upon the complexity and nature of the project and the density of reporting required.

3.4.2 However, effective time management cannot be achieved unless the planned sequence of events is monitored to reflect actual progress and the effect of intervening events is controlled. This cannot be managed effectively if the planned schedule is based upon a simple bar chart of activities. A logic-linked activity network is therefore the foundation upon which effective schedules and project controls are built.

3.4.3 Whichever scheduling technique is adopted, it must be anticipated that change will occur during the project timescale and that adjustments and amendments will need to be reflected within the time-model.

3.4.4 There are various ways of illustrating scheduling techniques, some of which are as follows:

■ bar charts

■ line-of-balance diagram

■ time-chainage diagram

■ arrow-diagram method

- precedence diagram
- linked bar chart

3.4.5 Bar charts

3.4.5.1   Bar charts have no logic. They are useful for illustrative purposes in high-level, low-density reporting but are of no value in the management of time in complex projects.

3.4.6 Line-of-balance diagram

3.4.6.1   The line-of-balance diagram (see Figure 2), which is commonly used to indicate the progression of resources through a multiplicity of areas, is a useful technique for coming to grips with the underlying philosophy of resource logic and critical-chain management. The line-of-balance technique has been proven to be useful in the management of linear projects (e.g. rail, road and pipeline projects).

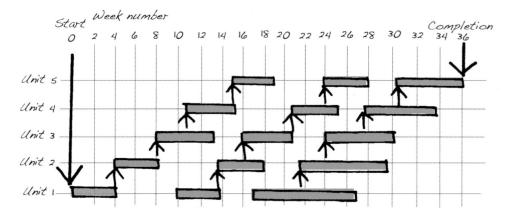

Figure 2 Typical line-of-balance diagram.

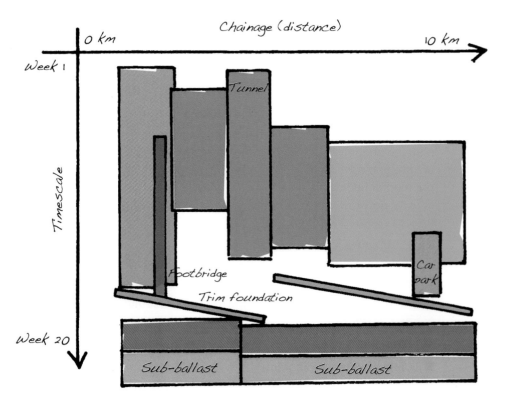

Figure 3 Typical time-chainage diagram.

It has a useful role to play in planning and resource scheduling of repetitive modular work (e.g. structural floors of multi-storey buildings).

3.4.7 **Time-chainage diagram**

3.4.7.1   Time-chainage diagrams are commonly used in connection with work of a linear nature to provide a graphical indication of both time and location of each work front (e.g. road, rail, tunnelling and pipeline operations), see Figure 3. Time-chainage diagrams have limited application in the construction of complex building projects.

3.4.7.2   Owing to the complex nature of a time-chainage diagram where multiple trades and work fronts may be included on a single diagram, the final output may be confusing and difficult to understand.

3.4.8 **Arrow-diagram method (ADM)**

3.4.8.1   This is the original critical-path-network technique. Before the advent of scheduling-software packages for personal computers, it was common to develop a schedule-activity network by hand, using arrow figuring where the activity was designated by the arrow and the logical interface of activities was represented by a node identifying the interface, as in Figure 4. The activity descriptions were originally contained in a separate document in order to reduce the physical size of the network.

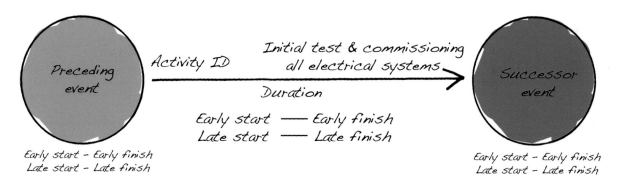

Figure 4 Typical activity diagram showing node–activity relationships.

3.4.8.2   An example of an ADM network identifying the activities on the network is provided in Figure 5.

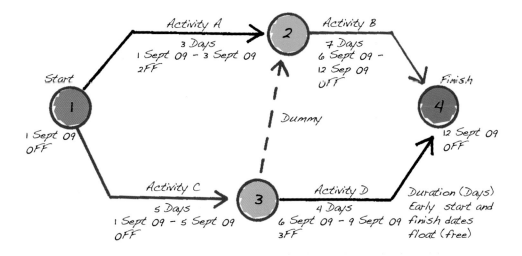

Figure 5 A four-activity ADM network.

3.4.8.3   In Figure 5, the 'dummy' illustrates a dependency between the completion of Activity C and the commencement of Activity B, which is not represented by any work activity. A similar technique must be adopted for leads and lags.

3.4.8.4   ADM networks are less prone to manipulation than PDM networks.

3.4.9   Precedence-diagram method (PDM)

3.4.9.1   This is the scheduling method adopted by most modern scheduling-software packages. In this technique, the activity is designated by a node containing the activity-related information, and the logical interface of activities is represented by an arrow identifying the interface. An example of a typical PDM node is illustrated in Figure 6.

3.4.9.2   In Figure 6, the activity is identified by an activity description and activity ID; in this figure the dates are illustrated as calendar dates and the duration and float identified in weeks, days or hours.

3.4.9.3   When linked together, a simple precedence figure looks like that illustrated in Figure 7.

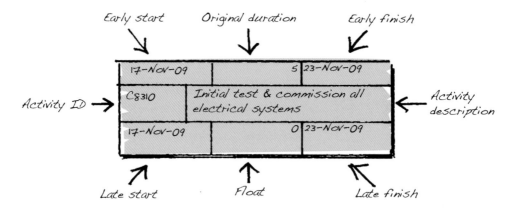

Figure 6 A precedence-diagram-method node.

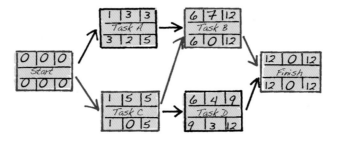

Figure 7 A four-activity PDM network.

3.4.10   Linked bar chart

3.4.10.1   This is another scheduling method offered by most modern scheduling-software packages. In this technique, the node is drawn as a bar, against a time scale in which the length of the bar is proportionate in length to its duration, and with its start and finish dates aligned with a calendar at the top and/or bottom of the figure. The activity-related information is contained in columns of data to the left of, and in line with, the activity bar. The logical interdependency of activities

3 Developing the time-model

Act ID	Description	Orig dur	Early start	Early finish	Total float	Aug														Sept							
						25	26	27	28	29	30	31	01	02	03	04	05	06	07	08	09	10	11	12	13	14	15	16
1000	Start	0	01 Sep 09		0																							
1010	Task A	3d	01 Sep 09	03 Sep 09	2d																							
1020	Task B	7d	06 Sep 09	12 Sep 09	0																							
1030	Task C	5d	01 Sep 09	05 Sep 09	0																							
1040	Task D	4d	06 Sep 09	09 Sep 09	3d																							
1050	Finish	0		12 Sep 09	0																							

Figure 8 A simple linked bar-chart network.

is represented by an arrow identifying the interface. An example of a typical linked bar chart, using the same data as is used in the PDM figure in Figure 7, is illustrated in Figure 8.

3.4.10.2   For most purposes, because of its infinite flexibility in report content and its ease of interpretation, this is the network-illustration method of choice for project reporting. However, as a scheduling method it has shortcomings in that:

■   the construction of a network as a linked bar chart tends to encourage the user to think in terms of lists and dates rather than in terms of logic and sequence;

■   the durations (represented as bars) tend to encourage the user to move the bars to 'paint the picture' which the user wishes to dictate, rather than to permit the software to calculate the sequence and dates; and

■   it tends to be more difficult, than with other scheduling methods, to make changes to the logic with the addition and/or deletion of activities and even to make changes of logic and sequence to the existing activities.

3.5   Resource planning and scheduling

3.5.1   **Resource planning**

3.5.1.1   Resources are labour, plant, money and materials. Space and time can also be viewed as resources. One or more of these will be more important than the others in planning different types of work; in most construction projects, labour will be the most common variable across most types of work, but in earthworks, for example, machine type and numbers as well as the production capability of the processing plant will be more relevant.

3.5.1.2   At low and medium density there will often be insufficient data available to make precise computations of durations by reference to productivity and resource alone. At these densities, it is quite often the case that activities have a mixture of resources and then other methods of estimation are therefore necessary. For example, estimation by reference to previous projects, experience, standard outputs and so on is acceptable at low and medium densities.

3.5.1.3   At high density, the planned duration of an activity is a function of the quantity of work, the productivity quotient and the quantity of the resource type to be deployed, in the formula:

$$\text{Duration} = \frac{\text{Quantity of work}}{\text{Productivity quotient} \times \text{Quantity of resource}}$$

3.5.1.4   The productivity quotient of a resource can vary according to the crew size adopted for the work, but the relationship between crew size and productivity is not linear. Reducing resources below optimum introduces inefficiencies, as does adding more resources above the optimum. Loss of efficiency needs to be balanced against space, cost, time and logistic considerations in order to determine the appropriate crew size for the work, and therefore the productivity quotient and duration of the activity.

3.5.1.5   The productivity quotient of a particular resource can be determined in a number of ways:

- published output rates

- historical data from other projects

- advice from specialists

- personal experience

- benchmarking

3.5.1.6   Resourcing the schedule validates the assessment of the durations adopted at lower densities. In addition, a resourced schedule permits:

- an understanding of work flow for gangs/trades. A steady work flow will deliver the most efficient plan;

- reassurance to other interested parties (internal or external) on the practicality of the time-model; and

- an understanding of any challenging resource requirements, for example, high-peak requirements, or lack of continuity for trades or gangs.

3.5.2   Resource scheduling

3.5.2.1   Resource scheduling is a time-intensive exercise and cannot be achieved without the participation of all those involved in the work. However, in order to demonstrate a reasonable, rational and valid high-density working schedule, work must be resource scheduled.

3.5.2.2   Resource scheduling is quite different from the technique of adding resources to a critical-path network for the purposes of calculating activity durations, or resource levelling carried out within the ADM/PDM logic network.

3.5.2.3   Although in building construction, work on site will generally follow activities of short duration in discrete areas, which are, to a great extent, subject to a critical path (and thus subject to critical-path network), time-modelling of complex projects will often include a number of other construction works, some of which will be in the nature of civil engineering and/or mechanical engineering and will not necessarily be subject to the same type of time control.

3.5.2.4   For example, in major land clearance, or cut and fill of land profiles, work may continue over a long period and may usefully follow some sequence (land will be dug out before depressions are filled), but not necessarily a sequence which is cogently linked from beginning to end of the operation as a whole.

3.5.2.5   Commonly, such a process may usefully be mapped out in what is referred to as a time-chainage diagram, which illustrates, in linear terms, where work is intended to be carried out. Although such figures may be logically linked to some extent, because they are drawn rather than calculated from a database, they do not commonly function as a time-model and other methods must usually be used to manage time and predict the consequences of change.

3.5.2.6   In such circumstances, where activities can be carried out over a long period in a multiplicity of areas, in any order (subject perhaps to limited sequences in any particular area), the time-model will commonly focus on the management of resources and productivity instead of critical-path sequences within the particular activity and critical-path sequences between sections of the activity and the interface with the remainder of the works.

3.5.2.7   It follows that such time-models are often subject to a high degree of revision to accommodate the preferential sequences that ultimately mature.

3.5.2.8   Where, for the purposes of high-density scheduling, the resources have been used to calculate the durations, the resourced schedules must be balanced prior to establishing the working schedule by moving non-critical activity excess demands to periods of available resource supply or vice versa.

3.5.2.9   The allocation of resources to activities to develop the high-density part of a realistic working schedule is a basic practice of network-based construction management.

3.5.2.10   Resource levelling is the process of balancing the peaks and troughs of resource demands. The process consists of two steps:

■   estimating the amount of resources needed by each schedule activity; and

■   using float values computed in the mathematical network analysis to schedule activities within the assumed availability of those resources. (Alternative criteria can be used to test several scheduling schemes.)

3.5.2.11   Not all resources are difficult to obtain (for example, resources which are readily available in abundance, locally, are unlikely to constrain performance) and where they are freely available they may usefully be planned to pace the project. The critical resources must thus be balanced first and then the impact on the other resources can be shown graphically by individual resource histograms.

3.5.2.12   Balancing the critical resources can be achieved with alternative criteria, using either early start or total float in a series of iterations, resulting in alternative performance schedules for consideration by the management team.

3.5.2.13   Wherever possible, trade teams should be scheduled to maximise continuity and to minimise alternating work and non-work periods. Where practicable, trade-related histograms can be used to help to illustrate how a succeeding trade's start could be delayed to achieve continuous performance.

3.6   Software considerations

3.6.1   General matters

3.6.1.1   At its lowest level the software available for scheduling may be no more than a drawing tool or, at its highest, a complex arrangement of customisable databases with a graphical front end. In order to be capable of producing a schedule which can perform as a time-model, the software must have an adequately functional database at its core. This is important because the software has to be capable of computing the consequences of change, whilst a drawing tool, which simply illustrates the decisions made by the drafter, cannot perform this core function.

3.6.1.2   No matter how high the quality of the software, it cannot produce a high-quality output of its own accord. Project scheduling software, even of the highest quality, will not ensure the competent management of time. The best that software can achieve is pointing the scheduler in the direction of good practice rather than encouraging picture painting.

3.6.1.3   Many software manufacturers provide extremely important and useful training. However, this form of training should not be confused with training in time management, nor should it be considered as a substitute for it. By analogy, many of us have experience of securing a good grounding on how MSWord works, but even with its spell checker and grammar checker the software will not guarantee that what is written is useful, technically accurate, or even intelligible.

3.6.1.4   Whilst, from time to time, every company considering software products will wish to take into consideration matters peculiar to themselves, or matters peculiar to the project upon which they wish to work, there are certain considerations that should transcend subjective preferences, and there are certain software attributes that are desirable for the purposes of competent time management.

3.6.1.5   Because software changes by the day, as 'new and improved'[3] products are brought to the market, those attributes that are desirable only for the purposes of time management, irrespective of whether they are currently available in any particular product, are listed in Appendix 2.

3.6.1.6   It is unhelpful if different parties to a particular project use different software because different products work in different ways and, even if given the same data, may produce different calculations from different algorithms. Accordingly, all parties to a project should use the same software and a departure should not be permitted.

3.6.1.7   Whilst getting to grips with unfamiliar software may be tedious, a competent scheduler who is provided with the recommended software training and some time to practise will be able to use the system and be sufficiently capable reasonably quickly. Unfamiliarity with scheduling software products should thus not be a serious consideration in product selection.

3.7   Schedule design

3.7.1   Introduction

3.7.1.1   The schedule is the means by which the strategy and sequencing of the full scope of work is to be accomplished. The scheduler must consider the type of schedule, its intended content, layout, appearance and the nature of any reports likely to be necessary before commencing scheduling.

3.7.1.2   The purpose of schedule design should thus be to set down the policy for rendering project plans and schedules as meaningful and transparent as possible.

3.7.2   The nature of the work

3.7.2.1   Work which can be carried out in defined sequences, over brief periods, requires a different approach to time management from work containing activities that may take several months to complete. The former can be managed by reference to the activity start, work in progress and completion date. However, the latter

---

[3] This unfortunately is a term often confused with more bells and whistles.

can only be managed from day to day by reference to the applied resources and productivity achieved.

3.7.2.2   Resource-based planning will thus be necessary wherever productivity is more likely to affect completion than logical sequence. Typically this can be applicable to such activities as earth moving in large-scale projects, piling on open sites, pipe welding on large process plants and other linear projects. In this form of time management, the unit productivity of each resource is interpolated as the works proceed to provide the data for calculating the time that the activity will take to complete.

3.7.3   **Schedule integration**

3.7.3.1   In most cases construction will involve a number of specialist contractors, subcontractors and suppliers. Design work may also be carried out by a number of design consultants, and some specialist contractors may also have design responsibilities.

3.7.3.2   The method of integration of the schedules of others and their transparency to the working schedule are important considerations in its design. The maintenance of incompatible and/or independent schedules by other contractors, subcontractors, suppliers or designers is inconsistent with effective time management.

3.7.4   **The time unit**

3.7.4.1   The time unit planned must be the same as the time unit against which production records can be kept.

3.7.4.2   For most building work, a time increment of a day is usually considered sufficient to be meaningful and anything more detailed is likely to be unmanageable.

3.7.4.3   However, some types of work, particularly those concerning facilities in use, may require short possessions to be planned by the hour and minute and, if relevant, the schedule design must permit this.

3.7.5   **The scheduling technique**

3.7.5.1   Critical-path-method-network diagrams are essential for illustrating the planned sequences, interrelationships and dependencies between activities and for demonstrating the logic of carrying out the works. CPM networks also provide the necessary model for predicting consequences and managing the effects of intervening events by executing 'what if' scenarios in the light of the intended logic. The robustness of the activity network is essential to subsequent analysis of the schedule. The principal alternatives are:

■   arrow-diagramming method

■   precedence-diagramming method

3.7.5.2   Depending upon the software product, the schedule may be configured to appear as one or more of the following:

■   an activity-on-arrow diagram

■   a precedence diagram

■   a linked bar chart

■   a bar chart

■   a data schedule

3 Developing the time-model

3.7.5.3 Because its content is most easily assimilated, for most low-density reporting processes a bar chart is likely to be the preferred display technique. A linked bar chart is also easy to understand and, depending upon the ability of the software product to vary the visual scale of the image, may provide a useful figure for logic tracing. However, a network produced by logic is much easier to follow than a bar chart because it is horizontal. The difficulty in reading a large bar-chart format is because of its verticality. For medium-density and high-density scheduling logic tracing, the ADM or PDM is usually necessary.

3.7.5.4 A data schedule will be necessary for productivity analysis and is usually essential for quality-assurance auditing of a CPM network.

3.7.6 Colours, fonts and graphics

3.7.6.1 Most scheduling-software packages will offer multiple formatting options. The temptation to use as many of these combinations as possible in any one chart should be avoided as the result may distract the reader from the data.

3.7.6.2 However, for clarity, care should be taken in the selection of fonts and colours so as to distinguish clearly between the structural elements of the schedule. Cartographers know that a map of any degree of complexity can be filled with only four colours without planes on a common border having the same colour. The same degree of clarity can be achieved on a schedule by limiting the number of colours used and avoiding multiple patterns.

3.7.6.3 In the selection of colours and patterns it should be borne in mind that, for some purposes, diagrams may need to be reproduced or transmitted in black and white.

3.7.7 The structure of the schedule

3.7.7.1 There are a number of structural matters that must be given consideration and upon which decisions need to be made before scheduling commences. Some of the key constraints that may limit scheduling options may include:

- project scope and objectives

- resources and labour, equipment and material constraints

- permits and licences

- utilities and third-party projects

- calendars

    Other decisions which will significantly impact the design of the schedule, but which are amenable to change, if necessary, are:

- work-breakdown structure

- activity-identification coding

- work-type definition

- density design

- contingencies

- activity-content coding

- activity-cost coding

- reporting

■ review revision and updating the schedule

■ impacting intervening events

3.7.8 Work-breakdown structure

3.7.8.1 A work-breakdown structure (WBS)[4] is a means of setting out the full project scope of work into manageable component parts by reference to a structured hierarchy.

3.7.8.2 The establishment of a WBS assists with control and visibility of project components and makes them into manageable elements of work, at high-density level, each element being assigned to specific resources.

3.7.8.3 The WBS must provide a hierarchical structure starting, at its highest level, with the project as a whole and, at the lowest level, ultimately identifying all work to be accomplished, for example, project, phases, areas, work packages and deliverables.

3.7.8.4 Each project will have its own idiosyncrasies and will necessitate the design of a workable template for the assignment of tasks and deliverables according to the selected WBS structure.

3.7.9 Activity-identifier codes

3.7.9.1 An activity-ID code is the unique unifying data identification which ties together the schedule, planning method statement and progress records. It is important that the structure for the naming of activity-ID codes is defined.

3.7.9.2 This usually involves determining a coding structure that is sufficiently flexible to provide identification by reference to a variety of criteria. For example, the activity code could be based upon:

■ the project and sub-project zonal configuration

■ the nature of the work to be scheduled

■ the contractual significance of the work

3.7.9.3 Some software products will automatically suggest an activity-identifying code and/or increment it according to selective data set up in the project definitions.

3.7.10 Work type

3.7.10.1 The work types which should always be identified are:

■ key dates, sectional completion dates and completion;

■ final dates for receipt of design information;

■ contractor's design and approval;

■ statutory approvals, permissions and works;

■ sample production and approval;

■ off-site fabrication;

■ equipment order, procurement period, delivery and installation;

■ access and egress to and from the site, and parts of the site;

---

4 For a definitive explanation of WBS, see US Military Handbook 881B, Australian Standard AS4817: 2006; and BS6079 – 1: 2002.

■ temporary works;

■ temporary plant order, installation, commissioning and decommissioning;

■ construction;

■ testing and commissioning of installations; and

■ handover process.

3.7.11 Schedule density

3.7.11.1   Unless the work is designed in its entirety and all subcontractors and specialists are appointed before any work commences, at the outset it is unlikely to be possible to plan the work in its entirety.

3.7.11.2   However, if time is to be managed effectively, the activities to be carried out, the resources to be applied and their expected productivity must be identified before work on the activity commences.

3.7.11.3   The density of the schedule may thus be expected to increase from that which is possible and necessary for feasibility purposes, as better and more certain information becomes available. A typical profile of schedule density against predictability is illustrated in Figure 9.

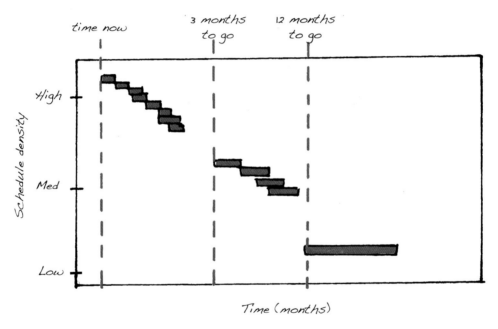

Figure 9 Graph of schedule density in relation to predictability.

3.7.11.4   The requirements of different densities of scheduling for different purposes must be taken into consideration at the schedule design stage and should be defined in the planning method statement.

3.7.11.5   A profile of an activity, as it varies from low density through medium to high density, is illustrated in Figure 10.

3.7.11.6   The relationship between low-density, medium-density and high-density parts of the schedule can be conveniently illustrated by taking a low-density activity and developing it through medium and high densities. The diagram in Figure 10, above, illustrates Activity B in low density taking 25 time units which in medium density is represented by Activities B1 to B4 and in high density by B11 to B48.

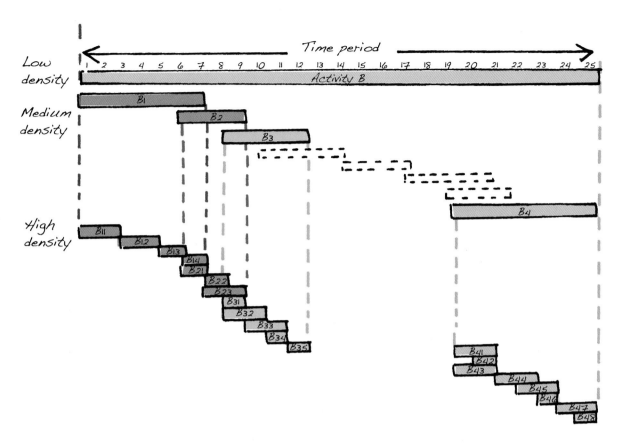

Figure 10 Illustration of schedule density.

3.7.12 Scheduling at low density

3.7.12.1 Low density is appropriate for work which is intended to take place 9 months or more after the schedule date. Depending upon the purpose for which the schedule is intended, tasks may reasonably be no more than the proposed duration of one building type, amongst others, or be trade grouped into such descriptions as 'mechanical and electrical services' and may conveniently be several months in duration.

3.7.12.2 Typically, standardised layouts must be prepared, usually as a bar chart, or linked bar chart, to illustrate such features of the works as:

■ the periods available for finance approval, licensing and permissions, design procurement and construction;

■ the order and sequence of construction of different buildings, site works and civil engineering.

3.7.13 Scheduling at medium density

3.7.13.1 Medium density is appropriate for work that is intended to take place between 3 and 9 months after the schedule date. At this stage the work should be designed in sufficient detail to be allocated to contractors or subcontractors for pricing.

3.7.13.2 Activities may reasonably be grouped into trade activities in locations of durations not exceeding 2 months.

3.7.13.3 Taking the same trade example, at this density the electrical services should be identified separately from the mechanical services, and the work to both services should be identified by area and zone.

3.7.14 Scheduling at high density

3.7.14.1   High-density scheduling is a prerequisite of work that is intended to take place in the short term, say, within 3 months after the schedule date; it is at high density that the work in progress will be recorded, monitored and reported upon.

3.7.14.2   At this stage the work should be designed in detail, the sequence and intended progress of the work clarified and the resources and productivity identified.

3.7.14.3   At this level the activity duration should be related to discrete tasks identified by a limited area and be no greater in duration than the period against which progress is reported.

3.7.14.4   Typically, standardised layouts must be prepared to illustrate such features of the works as:

- the detailed sequence to be followed by each resource for the identified period;

- the status of the work attributed to each resource at reporting periods;

- the relationship between resources and productivity planned and achieved;

- the relationship between costs incurred and productivity achieved; and

- the relationship between costs incurred and costs paid.

3.7.14.5   Whether reports are to be issued as electronic data files, or as hard copy will need to be established and, if in hard copy, the preferred paper size will constrain the reporting layout options available.

3.7.15 Calendars

3.7.15.1   Calendars have to be defined in order to establish the amount of working and non-working time against which an activity duration can be calculated.

3.7.15.2   Typically, calendars are established for the time generally available to the project, taking account of the working week, weekends, holiday periods and the like, and also for specific resource working hours (e.g. a resource may only be available to the project for certain days of the week, or restricted to certain hours of each day, or months of the year). It is also possible to establish calendars for exceptional periods of time (e.g. 24-hour working).

3.7.15.3   Most modern scheduling software comes with a selection of pre-set calendars which may often be used as a starting point from which to develop project-specific calendars.

3.7.15.4   There are a number of ways in which the effect of calendars may be illustrated on the network, but whichever method is selected, the chosen combination of display facilities should produce clarity.

3.7.15.5   For example, on a linked bar chart, the effect of calendars can be illustrated so that:

- non-working periods are illustrated as coloured vertical bands, in effect blanking the non-working days;

- selectively, illustration of some non-working periods, such as weekends, can be omitted in order to reduce visual clutter;

- the task bars can be shown in the foreground, or background, in relation to the non-working bands.

3.7.15.6 Some scheduling software will also permit the allocation of work-stoppage periods (sometimes known as exceptions) to which different colours and patterns may be applied. Again, the use of colours and patterns should be used only to achieve clarity.

3.7.16 Resources

3.7.16.1 Different scheduling software will deal with resources in different ways and it is important to recognise what the software will do and what it will not do before deciding how resources are applied to the schedule.

3.7.16.2 The resources that should be identified separately are all those categorised as:

■ separate contractors

■ subcontractors

■ consultants

■ clients

■ suppliers

■ plant

3.7.17 Permits and licences

3.7.17.1 The schedule should identify permissions which are required to be in place prior to work, or sections of work, being commenced or completed and should identify, separately, those to be obtained by the employer from those to be obtained by the contractor.

3.7.17.2 For each permit or licence there will be a period of time to be allowed for:

■ preparation of submittal

■ decision period

■ grant of permit or licence

3.7.18 Utilities and third-party projects

3.7.18.1 Independent projects which have to be carried out at the same time as the works, and upon which parts of the works are often dependent, create special problems. For example, supply authorities for water, gas, electricity, communications and the like require special consideration in the design of a schedule, and in relation to each supply utility, provision should be made for the identification of:

■ investigation of requirements

■ contract

■ grant of permit or licence

■ mobilisation

■ work period

■ highways

3.7.18.2 It should be borne in mind that work in connection with investigations may be dependent upon prior agreement for diversions.

3.7.19 Contingencies

3.7.19.1 Typically, there will be many risks to be considered, some of which may require a contingency period.

3.7.19.2 Contingency periods should be designed[5] to be identified separately for both the employer's and the contractor's risks and for those risks which are related to:

■ an activity or chain of activities;

■ a contractor, subcontractor, supplier or other resource;

■ an access or egress date, or date of possession, or relinquishment of possession;

■ the works, any defined section, and any part of the works.

3.7.20 Activity-content coding

3.7.20.1 Except in unusual circumstances, the schedule will be reviewed and revised from time to time. Events will occur which will need to be taken into account in the forward planning of the work and progress achieved will be different from the productivity expected. In order to manage the time implications of these departures, an activity-coding structure is important and can usefully assist in filtering operations where a single view of individual trades or disciplines is required. Consideration should be given to such matters as:

■ changes to schedule density;

■ change of calendar;

■ resource allocations;

■ creating or amending hammocks;

■ inserting resource-driven links; and

■ providing a short-term schedule of work for each trade or discipline.

3.7.20.2 Different scheduling software will deal with activity coding in different ways and it is important to understand how the scheduling software handles activity codes before attempting to apply them.

3.7.20.3 Activity codes should identify the various attributes of the schedule as fields, the values of which will facilitate organisational changes, and facilitate filtering of important parts of the schedule. Activity coding is thus used to provide:

■ visual impact and identification of an activity group on the schedule; and

■ an aid to filtering, linking and organising activities into suitable 'views' for quality assurance, auditing and reporting purposes.

3.7.20.4 Most current scheduling software products will offer the scheduler a multiple choice of formatting options for the presentation of assigned activity codes. In so far as the software defaults to using particular colours, or patterns to indicate critical and non-critical activities, they must not be used for the indication of other fields, or values in other fields.

3.7.20.5 If the scheduling software supports the allocation of multiple tasks to a single bar for repetitive tasks such as high-rise structures, road operations, pipelines, or procurement stages, then the judicious use of activity codes can assist in the presentation of tasks up or down the project, visually similar to the picture generated by a time-chainage schedule.

---

[5] See 'Risk and contingencies' in paragraphs 3.8.54 to 3.8.57.3.

3.7.21 **Activity-cost coding**

3.7.21.1   Cost coding works in a similar way to activity-content coding in that given particular cost budgets, or fields, values can be attributed to them in relation to particular resources or activities.

3.7.21.2   An effective cost-coding system is an integral aspect of project control. If adequately configured to do so, it can be used effectively in calculating interim costs from interim progress updates and can help to avoid potential misalignment of anticipated completion values.

3.7.21.3   Cost codes also facilitate an alignment between cost accounting and management accounting so that the financial reporting tends to be more accurate.

3.7.21.4   A cost-coding system does not need to be complicated in order to produce valuable data and should be designed to meet the complexity level of the other required project controls. The codes and structures should be considered in much the same manner as the establishment of the WBS and activity codes. Their application should be thought through in advance of their application onto the schedule, such that the desired level of analysis and reporting can be achieved.

3.7.22 **Schedule reporting**

3.7.22.1   In complex projects it is impracticable to use the whole of the schedule at any one time in its detail. For effective reporting it should be summarised to different degrees of summarisation for differing purposes (see Figure 11, below). Most project-scheduling software packages facilitate this hierarchical structuring by virtue of a summarisation or roll-up facility.

3.7.22.2   Typically, standardised layouts must be prepared to illustrate such features of the works as:

■   the intended sequence and timing of specific trades or processes;

■   information release dates, submittal and approval dates;

■   the effect of progress achieved from time to time on each completion, sectional completion or key date; and

■   the effect of any one or more intervening events on each completion, sectional completion or key date.

3.7.22.3   The following provides an outline of the usual reporting levels for which the design should allow at the outset:

■   **Level-1 report** – Also called an executive-summary report. This represents major milestones in the schedule; it highlights major project activities, milestones and key deliverables for the whole project. It is used to summarise the schedule in low-density reports and other documents when a more detailed schedule is not required and may reasonably be prepared as a bar chart.

■   **Level-2 report** – Also called a senior-management report. Maintained as a summary of the level-3 schedule, it depicts the overall project broken down into its major components by area and is used for low-density management reporting.

■   **Level-3 report** – Also called a project manager's report. Maintained as a summary of the level-4 schedule for reporting status to senior management and to report monthly status to the employer. It is usually in medium density.

3 Developing the time-model

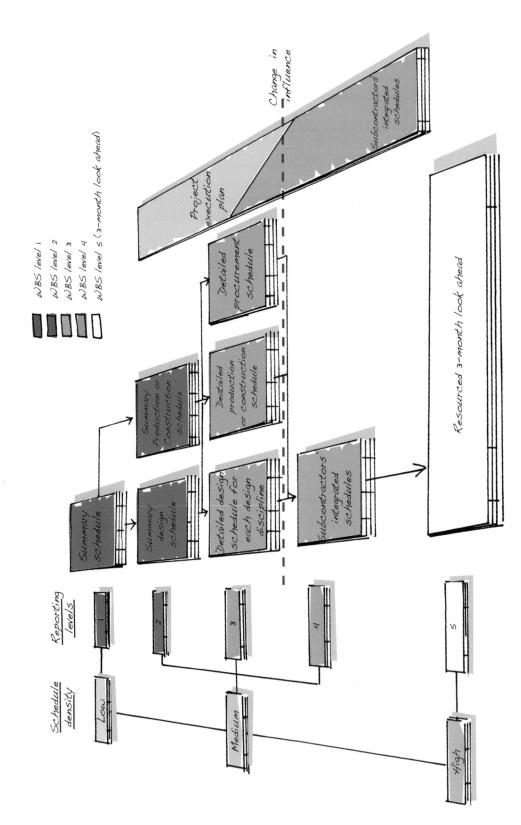

Figure 11 WBS levels and schedule density.

■ **Level-4 report** – Also called a section manager's report. Level 4 is the detailed working-level schedules, where each schedule is an expansion of part of a level-3 schedule and is established within the integrated project schedule. This is the working-level schedule displaying the activities to be accomplished by the project workforce in medium density. The dates generated by these activities represent the anticipated start and completion of work required to complete the project.

■ **Level-5 report** – Also called a short-term, look-ahead report. It illustrates the further breakdown of the activities of a level-4 schedule into a high density, short-term schedule used to map out the detailed tasks needed to coordinate day-to-day work by resources allocated to specific areas of the project.

3.8 Schedule preparation

3.8.1 Work-breakdown structure

3.8.1.1 The first step in the preparation of a schedule is to define and implement a work-breakdown structure. A simple project WBS based upon zones of activity and stages of work is demonstrated in Figure 12.

3.8.1.2 An alternative WBS hierarchy illustrating a relationship of resources is illustrated in Figure 13, below.

3.8.1.3 Although the WBS may be amended at a later stage in the project, the resultant change may cause confusion with interrelated and associated controls, for example, cost systems established against the same structure, progress measurements and so on.

3.8.1.4 Accordingly, a clear, well-thought-out WBS must be prepared at the commencement of the project. Ultimately the WBS should result in discrete work packages of 'deliverables' that define the project-work scope and encompass allocated measurable periods of time and resources.

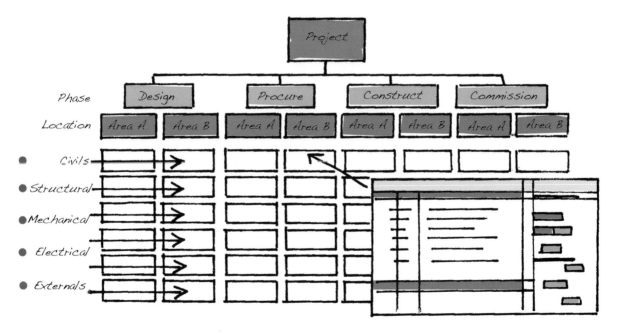

Figure 12 An example of a functional project WBS.

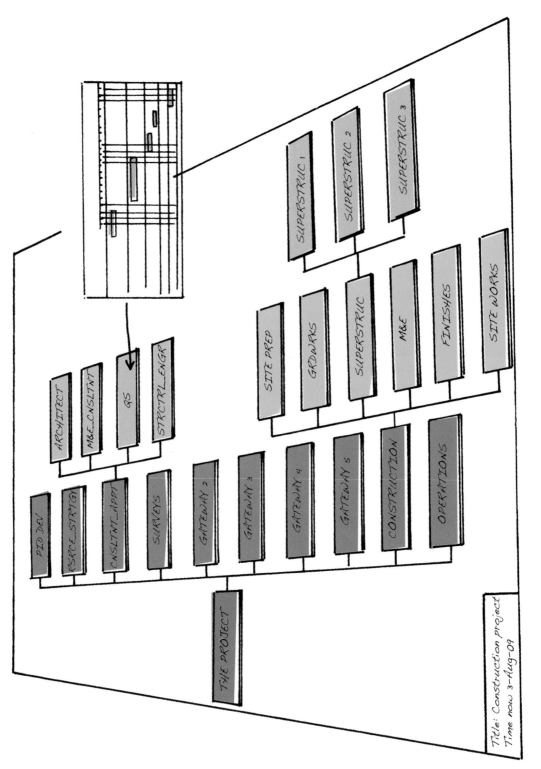

Figure 13 A typical WBS showing work-packages/professions/trades.

3.8.1.5   A heading should be created for project change so that the work involved in accommodating intervening events can be identified and any effect on the schedule more easily managed.

3.8.1.6   Once the template for the WBS is established, a range of project controls can be assigned to each of the work packages. These controls typically include the assignation of a responsible person for the delivery of the work package, key dates to be met within that package, a budget and an indication of key deliverables.

3.8.1.7   The WBS may also be integrated with other organisational structures such as the cost-breakdown structure or the organisation-breakdown structure (OBS) in a matrix fashion, to provide an alignment of work packages with associated budgets/costs, or organisation responsibilities. Care should be taken, however, in the development of such a framework, as it is all too easy to confuse the simplistic intent of this tool (which is to provide clarity of work, deliverability, timing and control) with an organisational structure which is too complex or unwieldy.

3.8.1.8   An integrated WBS, CBS and OBS is shown in Figure 14.

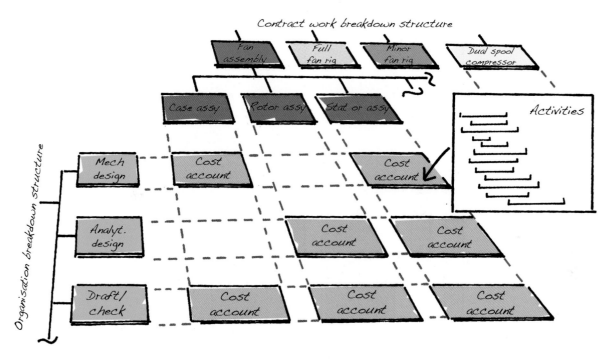

Figure 14 An integrated WBS, CBS and OBS.

3.8.2   Activity-identifier code

3.8.2.1   The activity ID should be capable of being broken down into related sub-activities, from low density to medium density and to high density, as the work progresses, without loss of an audit trail through the development of the activity detail. So, taking the same example as for description, the activity IDs could be constructed as illustrated in Figure 15, below.

3.8.2.2   In Figure 15, the descriptions have been kept simple for clarity but see Figure 16 for a breakdown of a more detailed approach to activity descriptions.

3.8.2.3   In this example, the letter Z has been reserved for a code-identifier column which has no data and, where there are no more than 25 buildings and no more than 25 zones, a single letter has been used to identify each.

3 Developing the time-model

Location Zone	Area	Section	Item	Description	Activity ID
Low density					
B	A			Substructures	ABAZZ00010
Medium density					
B	A	A		Excavations	ABABAZ00010
B	A	B		Piling	ABABZ00010
B	A	C		Ground beams	ABACZ00010
B	A	D		Floor slabs	ABADZ00010
High density					
B	A	C	A	Ground beams	ABACZ00010
B	A	C	A	Formwork	ABACA00010
B	A	C	B	Reinforcement	ABACB00010
B	A	C	C	Placing concrete	ABACC00010
B	A	C	D	Curing	ABACD00010
B	A	C	E	Strike formwork	ABACE00010
B	A	C	F	Backfill	ABACF00010

Figure 15 Example of activity-ID-coding structure.

Location	Zone	Area	Section	Item	Detailed activity description	Activity ID
Low Density						
A	B				Substructures inBld.A, ZN.B, Ar.A	ABAZZ00010
Medium density						
A	B	A	A		Excavations, Bld.A, Zn.B, Ar.A, Sec.A	ABAAZ00010
A	B	A	B		Piling, Bld.A, Zn.B, Ar.A, Sec.B	ABABZ00010
A	B	A	C		Ground beams, Bld.A, Zn.B, Ar.A, Sec.C	ABACZ00010
A	B	A	D		Floor slabs,Bld.A, Zn.B, Ar.A, Sec.D	ABADZ00010
High density						
A	B	A	C		Ground beams,Bld.A, Zn.B, Ar.A, Sec.C	ABACZ00010
A	B	A	C	A	Formwork to gnd bms,Bld.A, Zn.B, Ar.A, Sec.C	ABACA00010
A	B	A	C	B	Reinforcement to gnd bms.Bld.A, Zn.B, Ar.A, Sec.C	ABACB00010
A	B	A	C	C	Placing concrete to gnd bms.Bld.A, Zn.B, Ar.A, Sec.C	ABACC00010
A	B	A	C	D	Curing to gnd bms.Bld.A, Zn.B, Ar.A, Sec.C	ABACD00010
A	B	A	C	E	Strike formwork to gnd bms.Bld.A, Zn.B, Ar.A, Sec.C	ABACE00010
A	B	A	C	F	Backfill to gnd bms.Bld.A, Zn.B, Ar.A, Sec.C	ABACF00010

Figure 16 Unique activity descriptions.

3 Developing the time-model

3.8.2.4   All the work to be coded occurs in Building A and is in Zone B of Building A; therefore, no matter at what degree of density, all activity IDs commence with the two letters AB.

3.8.2.5   At low density, all we have is a single bar called 'substructures', which in this example is called Area A. Accordingly, the activity commences ABA.

3.8.2.6   At medium density, the 'substructures' are broken down into 'excavations', 'piling', 'ground beams' and 'floor slabs', each of which is given an identifier letter so that in this example 'ground beams' is called Section C. Accordingly, the activity commences ABAC.

3.8.2.7   At high density, the 'ground beams' are broken down into 'formwork', 'reinforcement', 'placing concrete', 'curing', 'strike formwork' and 'backfill', each of which is given an identifier letter so that in this example 'reinforcement' is called Item B. Accordingly, the activity commences ABACB.

3.8.2.8   The trailing numbers start at 00010 and would normally progress in tens to leave room for any other activities needed at a later date to be inserted within the bracket of 10 (so as not to break the numerical sequence). Thus, the next breakdown of the 'reinforcement' activities might be 00020, 00030 and so on, and if at a later date another reinforcement activity were to be needed between 00020 and 00030, it would have the activity ID of ABACB00025.

3.8.2.9   Alternatively, a simpler form of activity-ID coding may be adopted in some circumstances, whereby in the development schedule an activity ID is assigned depicting only the phase and zone of operation, for example, DA1000 = design, area A, activity 1000. This may then be developed into a working schedule by adding further breakdown as DA1000.01 to 99 and so on.

3.8.2.10   The importance here is not the complexity or simplicity of the coding formula but that it creates a meaningful activity ID which is suitable for its purpose in the circumstances and can be broken down into detail as the schedule density requires.

3.8.3   Activity description

3.8.3.1   Activity descriptions capture the essence of the intended project deliverables. Given that they are often truncated summaries of the fuller descriptions contained elsewhere, for example, in the employer's requirements, it is important that whatever density of schedule is adopted from time to time, the activity description is clear and unambiguous.

3.8.3.2   Most software products have a limited field for the activity description. Thus, apart from configuring a unique identifying code for each activity, it is important to establish a formula for the naming of each activity so that the descriptive content of each is unique and unambiguous.

3.8.3.3   Depending upon the purpose of the schedule, at feasibility stage the activity description may be so coarse as to facilitate no more than an intended duration, from start to finish, of each building within a group. On the other hand, at construction level, significantly greater detail will be required for time management. As the density of the schedule increases, so the clarity and unique quality of the description becomes more important.

3.8.3.4   However, whatever the density of the schedule, the activity description must be clear, concise and fit for its purpose.

3.8.4   Descriptions at low density

3.8.4.1   Initially, the schedule may be generated at a time in the project when the scope of work is not fully defined. Accordingly, for low-density schedules an activity description which facilitates later sub-division will be required (e.g. substructures).

3.8.5 Descriptions at medium density

3.8.5.1   It is important to note that the early construction-related schedule will be submitted to fulfil contractual obligations and it is of paramount importance, therefore, that all elements of the work are adequately described.

3.8.5.2   It is worth keeping the activity description at the early stage of development at a reasonably high level in order to maintain a degree of scheduling flexibility once the scope of work is further defined and detailed (e.g. excavations, piling, ground beams, floor slabs).

3.8.6 Descriptions at high density

3.8.6.1   The activity description must clearly identify, unambiguously, exactly what work is to be done and where it is to be carried out. Without such clarity, it will be impossible to accurately record the progress of it, for example:

Ground beams:

- formwork

- reinforcement

- placing concrete

- curing

- strike formwork

- backfill

3.8.6.2   An example of how the increasing density of activity description works in practice is given in Figure 16, above, which, by using the same example data as in Figure 15, above, illustrates how the descriptions are nested but rendered unique by reference to the data identifying such things as Location, Zone, Area, Section and Item.

3.8.7 Ascertaining activity durations

3.8.7.1   Depending upon the density of the schedule, the purpose for which it is to be used and the information available, an activity duration can be derived from:

- experience

- industry standards

- benchmarking

- comparison with other projects

- calculation from resources

- specification

3.8.7.2   One of the most important considerations in the development of the content of an activity network is whether the duration of any activity is to be determined inductively or deductively. In other words, the scheduler must determine whether the duration of any activity is to be determined:

- empirically, by setting down the date it will start and finish, assuming that sufficient resources can be made available to perform to that standard; or

■ quantitatively, by identifying the quantity of work to be performed, the resources which reasonably can be allocated and their rate of performance.

3.8.7.3 Most scheduling software will offer alternative activity rules for dealing with duration. The default is usually to give priority to the allocated duration in preference to any duration deduced as a result of applied logic, or resources although, normally, this may be changed to give priority to either of the other data categories when inconsistent with the allocated duration.

3.8.7.4 When resource allocation and designated productivity are identified as determining factors in the calculation of an activity duration, the duration will be calculated by reference to the given resource data.

3.8.7.5 At medium and low density, when activity duration is identified as to be calculated by reference to the logic of the schedule, for example, by extending an activity beyond its given duration to be consistent with finish-to-finish logic, it is usually referred to as a 'stretched' or non-contiguous activity. Scheduling software may permit this selection according to particular activities, or only in relation to the schedule as a whole.

3.8.8 Durations at low density

3.8.8.1 Durations will often be derived by assessment from the experience of those associated with the project.

3.8.8.2 Through experience it may be known that the erection of steelwork with a single gang of men can proceed at 'x' tonnes per hour, therefore 'y' tonnes of steel will be erected in 'y' divided by 'x' hours (e.g. 500 tonnes of steel (y) at 5 tonnes per hour (x) = 500 divided by 5 = 100 hours).

3.8.8.3 At this level much of the work may not have been fully specified and hence the definition of durations by reference to empirical data, with a high degree of formulaic contingency, will often be appropriate.

3.8.9 Durations at medium density

3.8.9.1 At this level the work can be expected to have been designed in detail and a higher degree of science should be applied to calculating the reasonable durations of activities.

3.8.9.2 At medium density there should be no approximation based upon an expected duration formula.

3.8.9.3 Activity durations will ordinarily be derived from one or more of the following:

■ the specified standards of performance (e.g. curing times, specified provisional and contingency periods, and similar provisions);

■ calculation by reference to quantities, approximate quantities, notional resources and standard production rates;

■ assessment by experience;

■ specialist subcontractor and supplier production data; and

■ historical production data.

3.8.10 Durations at high density

3.8.10.1 This density governs work intended to be carried out in the next three months, and all information concerning the designed works and resources should be available.

3.8.10.2 There should be no assessment by experience alone at this stage although experience will play a large part in determining whether planned production rates and resources are reasonable.

3.8.10.3 Activity durations will ordinarily be derived from one or more of the following:

■ the specified standards of performance (e.g. curing times, specified provisional and contingency periods, and similar provisions);

■ calculation by reference to quantities, approximate quantities and production rates;

■ specialist subcontractor and supplier production data;

■ historical production data; and

■ benchmarking.

3.8.10.4 In connection with the effects on durations of likely productivity, the following should also be considered:

■ physical working conditions;

■ safety requirements and labour agreements that may be in effect;

■ site downtime;

■ seasonal weather fluctuations;

■ seasonally related activities; and

■ resources for seasonal work.

3.8.11 Experience

3.8.11.1 Duration – estimating by experience in relation to one or more activities is likely to be essential in low-density and, to some extent, in medium-density schedules.

3.8.11.2 Construction trade professionals and managers at the workface will tend to know from their own experience how long it is likely to take to perform a particular activity in ordinary day-to-day working conditions with a given gang size. They are also likely to be able to give some guidance as to the likely effect that restrictive conditions may have on their productivity.

3.8.11.3 In the absence of objective standards to work to, the scheduler will have to fall back on their own experience. However, one technique still available is to gauge a range of possible durations from alternative sources and then to apply these in the following manner:

■ the optimistic duration – the shortest time in which the activity can reasonably be completed;

■ the most likely duration – the completion time having the highest probability;

■ the pessimistic duration – the longest time that an activity may reasonably require.

From a collation of the above, the following formula will give a weighted expected duration:

$$\text{Expected activity duration} = \frac{\text{Optimistic} + (4 \times \text{Most likely}) + \text{Pessimistic}}{6}$$

3 Developing the time-model

3.8.12 Industry standards

3.8.12.1 Resource and productivity data should obviously be as realistic as possible for the prevailing circumstances and work type. The data can be obtained from a number of sources but from wherever it is obtained, its source should be recorded in the planning method statement.

3.8.12.2 A schedule of currently available sources is set out in Appendix 4.

3.8.13 Benchmarking

3.8.13.1 With repetitive operations, a work study should be carried out to establish the performance baseline. If there are 500 bedrooms to be fitted out, the time it will take to fit out one bedroom will be an extremely important consideration: any error will be multiplied 500 times. Therefore, it will be very important to establish the correct duration for the first one by carrying out a work study. Similar considerations may be applied to the structural lifts of a tower block, piling, pipeline being laid, and, indeed, on a micro-level to any repeatable activity such as fitting a door. Thus, checking the exercise by a benchmark study, in sample, may be an extremely valuable exercise, and that is especially so if changes are later made which incur time-related and disruption-related costs.

3.8.13.2 Where benchmarked productivity data is available from the production data of similar work previously executed by the same contractor, this should be taken in priority over any other data which is not project-specific.

3.8.14 Comparison with other projects

3.8.14.1 In the absence of more objective data from which to work, taking as a baseline the known time from other projects of like content is often the preferred method of allocating time in low-density, higher-risk schedules in which there is a high degree of speculative time allocation.

3.8.14.2 In low-density and medium-density schedules, it may be prudent to engage with those specialist subcontractors who may be or have been appointed in order to accommodate experience of their speciality.

3.8.14.3 However, using as-built data from other projects is a technique of limited application in medium-density schedules in which the majority of design will have been completed and hence a more project-specific method of estimating activity durations will be appropriate.

3.8.14.4 The allocation of durations by reference to other projects of like kind has no application in high-density, short-term scheduling.

3.8.15 Calculation from resources and work content

3.8.15.1 The relevant resources are:

- labour
- machines
- money
- materials
- space

3.8.15.2 However, the order of priority of importance attributed to these resources will differ from time to time according to the type of work from project to project.

For example, machine type and numbers, as well as the production capability of the processing plant, will be the driving factors in earthworks, but in most construction projects, labour is likely to be the driving resource for most types of work.

3.8.15.3   Four methods of resource scheduling are usually available and it is useful if the scheduling software is able to calculate the schedule using all of them. Of particular importance is the facility for scenario modelling. The four methods are:

■   the timeline and critical path determined through time-critical resource levelling (the project end date will not change and will show overloading of resources);

■   resource-critical resource levelling (the project end date will change reflecting the time required to complete the project with the currently allocated resources);

■   resource analysis using the activities' early start and finish dates; and

■   resource analysis using the activities' late start and finish dates.

3.8.15.4   In the high density part of the schedule, durations should be calculated by reference to resources and productivity in high-density schedules.

3.8.15.5   At high density, the duration of an activity is to be a function of the quantity of work and the productivity quotient relevant to the resource allocated. This method of calculating durations may also be available to some activities in medium-density schedules, but in low-density schedules there will often be insufficient data available to produce meaningful calculations.

3.8.15.6   Once the allocation of resources has been made and activity durations recalculated according to the productivity quotients, it may be found that the resultant schedule assumes unnecessary float with discontinuous trade activities. In order to achieve a higher degree of continuity, at the expense of float, the resource allocations can be adjusted. This process is called resource levelling.

3.8.15.7   The object of resource levelling is to smooth out requirements for labour and plant resources so that, wherever possible, peaks and troughs, and hence predictable loss of productivity, are avoided. The purpose is to avoid arriving at a schedule which, for example, requires 100 men and 20 machines one week, none at all the following week, and 50 men and 10 machines the week after that.

3.8.15.8   Levelling is achieved by adjusting the timing and/or duration of activities in order to even out the resource requirements. Essentially, apart from changing the network logic, there are two methods by which peaks and troughs in resource requirements can be avoided:

■   reducing the resources to fill the available time; and

■   increasing the durations of non-critical activities.

3.8.15.9   Both these methods make use of what would otherwise be available float overriding early and late start dates of levelled activities and rescheduling them to dates when the smoothed resources become available.

3.8.15.10   Resourced schedules can provide resource histograms and cumulative graphs to assist those who have difficulty in reading raw data to review, easily and rapidly, at least one resource, or element of the schedule at a time.

3 Developing the time-model

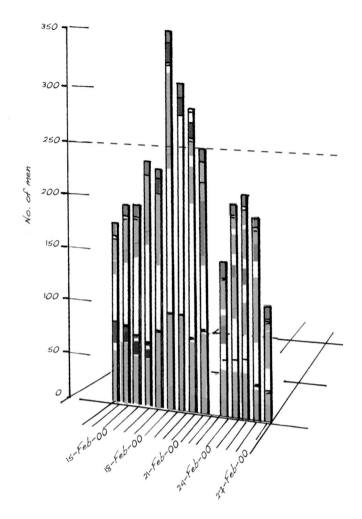

Figure 17 Multiple resources prior to levelling.

3.8.15.11   Figure 17 illustrates different resources in different colours prior to rescheduling and levelling. By review of an illustration such as this, the scheduler can examine the resources' use of the trade, shown in grey/white above the dotted line and consider the practicality of achieving the levels shown.

3.8.16   Specification

3.8.16.1   Notional periods for provisional work, work relative to prime cost sums and other contingencies should be defined and specified by the project team for inclusion in low-density schedules and in medium-density schedules in which the previously unknown work still remains undetailed.

3.8.16.2   Schedule density will be the deciding factor as to what is an appropriate and reasonable method for determining the schedule durations.

3.8.17   Calendars

3.8.17.1   The project calendar will set out:

■   the normal working day; and

■   working time within working days to accommodate the number and timing of hours per shift, and shifts per day.

3.8.17.2   The calendar will also identify exclusions from the normal working day to accommodate, for example:

- weekends

- statutory holidays

- religious holidays

- other industry-recognised holidays

- local holidays etc.

3.8.17.3   Separate working periods may need to be defined to take account of the working periods of, for example:

- specialist contractors

- subcontractors

- suppliers

3.8.17.4   For any resource, working periods may also need to be defined to take account of such things as, for example:

- limited possessions

- weather cycles to which the resource is sensitive

- planned overtime

- power outages

- environmental considerations[6]

- earthworks in winter

- temporary traffic diversions

- plant maintenance downtime

3.8.17.5   Multiple calendars should be set up to account for anticipated work stoppages over and above weekends and public holidays (such as inclement weather, railway possessions and power outages), embargo periods (for such things as earthworks in winter, river crossings in the fish-breeding season), road closures' access availability, multiple-shift working and planned plant maintenance downtime, and the like.

3.8.17.6   However, because of the difficulty in identification and traceability, calendars must not be used to allow for contingency periods by identifying, for example, what would otherwise be working days as notional non-working days to allow for potentially adverse weather conditions. The correct way to make such allocations is as contingency activity periods which can plainly be seen and audited.

3.8.17.7   Calendars for activities which are not related to the site-operating calendar, but to calendar days of 24 hours each, will also be required to accommodate such periods as:

- submittals and approval consideration periods;

- procurement and delivery periods;

- mobilisation periods;

- concrete curing time.

3.8.17.8   Depending upon their purpose, such calendars may or may not recognise religious, industry-recognised and public holidays, and weekends, and each must be considered on its merits.

---

[6] For example, the effect on protected species and breeding seasons.

3.8.17.9 Project-scheduling software products do not all deal with multiple calendars in the same manner. Accordingly, it is strongly advised that calendar allocations are kept as simple as possible in order to gain a meaningful understanding of their effect on time. It should be borne in mind that the effect of the allocation of multiple calendars will be dictated by the scheduling algorithm adopted by the scheduling software, and for any given combination, different software may produce different effects on the calculation of criticality. Accordingly, it is important that the scheduler and the management team understand how the chosen software uses calendars in relation to both activities and resources.

3.8.17.10 Resource calendars are usually established by specific, named calendar definitions, or by defining the actual date changes in each particular specified resource availability. These may be allocated to the resource in the resource definition, but the way calendars can be allocated to resources is very much dependent upon the characteristics of the software being used.

3.8.17.11 Resource calendars will usually override the default activity calendar when rescheduling so care should be taken to allocate non-working periods correctly when identifying the working time of a particular resource.

3.8.17.12 Apart from identifying working days, the calendars must identify working hours within the working days. These are referred to as 'work patterns'. Work patterns are the series of working and non-working times in any single working day. Only one work pattern can be assigned to any single working day for each activity, or resource. At the extremes, some scheduling software will allow the work-pattern time to be allocated in minutes and seconds, others may be coarser and limit the options to nothing more precise than half-days. Depending upon the scheduling software, work patterns may be allocated to apply to a year at a time, in others the options may be limited to weeks or months at a time.

3.8.17.13 Although it is technically possible to incorporate a night shift into what is otherwise a day-shift work pattern, in some software products some facilities may be adversely affected. Thus, if a night shift is needed then a different calendar should be allocated for it.

3.8.17.14 Different work patterns can be assigned to particular seasons in the same calendar, for example:

■ daylight-saving winter periods in hospitable working conditions;

■ extreme weather conditions in harsh environments where working in the middle of the day is either the only possibility or, alternatively, impracticable.

3.8.17.15 Work patterns must not be used to identify contingency periods by identifying, for example, what would otherwise be working hours as notional non-working hours to allow for potential lost productivity. The correct way to make such allocations is as contingency activity periods which can plainly be seen and audited from time to time.

3.8.17.16 The possibility of increasing working hours beyond the norm in order to shorten durations should be avoided. Increased working hours for other than a brief period will always result in a loss of productivity, and unless productivity output and increased risk of rework is also taken into consideration, the apparent shortening of the duration is likely to be misleading. Care should be taken to ensure that the allocated working patterns are consistent with local health and safety requirements.

3.8.18 Calendars at low density

3.8.18.1   Initially, calendars may be limited to those identifying the activity-related working week, together with statutory and other holidays.

3.8.19 Calendars at medium density

3.8.19.1   At this level all relevant activity-related calendars and the major resource-related calendars should be in place.

3.8.20 Calendars at high density

3.8.20.1   By the time the work is about to start on site, all resource calendars must be established and allocated to the appropriate resources to be employed.

3.8.21 Activity-content codes

3.8.21.1   Project-scheduling software is based upon a database application containing the facility for the creation of many different data fields and values within those fields. One type of field is known as the 'activity code'. By the attribution of codes and values to a group of activities, they can be searched, filtered and displayed as a discrete group to facilitate auditing and quality assurance, review, revision and reporting.

3.8.21.2   Typically, the schedule may be organised either by means of the WBS or by the use of activity codes. However, in so far as the WBS is a relatively static structure, because they are attributed for particular purposes, activity coding provides an infinitely variable structure against which the schedule may be rearranged.

3.8.21.3   Examples of typical descriptive codes and values that may be attributed are illustrated in Figure 18.

Descriptive Fields	Descriptive Values
Location	Building 1, Building 2, River crossing
Area	Basement, Ground, 1st Floor, 2nd Floor
Zone	Gridlines, A1-G5, Gridlines A6-G10, Gridlines A11-G15
Department	Design, Purchasing, Information systems, Construction
Responsibility	Project director-Alan Cappins, Chief architect-Dave Schmit, Director of purchasing-Ellen Ulrich
Phase	Conceptual design, submittals and approvals, shop drawings, procurement, construction, testing
Section	Foundations, structural frame, cladding, installations
Events	A1, A12, CVC4, Claim1, Claim2, Claim3

Figure 18 Example of descriptive activity-content codes and values.

3.8.21.4   It is preferable to set up the activity-code groups that are most likely to be applicable to the schedule, in advance of setting down any tasks to which the fields may be applied. Due account should be taken of 'common' code groups that may be applicable to an organisation, or to a project such that duplication of code groups is avoided within other areas of the enterprise.

3.8.21.5   Activity-content codes and values may be added at any time during the planning process. However, once they have been attributed to activities, the subsequent amendment of the order in which the codes or values were created, or their structure, may cause significant difficulties and should be avoided.

3.8.22 Cost codes

3.8.22.1   Cost codes and structures are used in much the same manner as activity codes and work-breakdown structures in that they are applied to a schedule for the effective subdivision of budgets and cost collation such that each cost package may be monitored and controlled.

3.8.22.2   The facility of cost codes arranged in a structure aligned to the WBS will present significant advantages to the project team. For example, the project cost engineer or quantity surveyor can be expected to predict the cost estimate to complete for any cost package based upon the level of progress, cash-flow forecast and productivity achieved. Most software packages facilitate reporting against this data to provide a dynamic and integrated view of project progress related to earned value against predicted cost.

3.8.22.3   The application of costs to activities and resources facilitates the alignment of project controls between the project planner, project scheduler and the cost engineer, or quantity surveyor. However, the degree of detail will largely depend upon the density of the schedule.

3.8.23   Cost coding at low density

3.8.23.1   At this level of density, cost data may reasonably be represented by major budgetary groups and/or predicted payment milestones in the schedule. Given this level of schedule density, it cannot reasonably be expected to contain other than a headline view of budgets and will normally contain a high level of contingency.

3.8.24   Cost coding at medium density

3.8.24.1   At this level, all relevant activities and many of the resources may be established. Cost-coding structures at this level are more often assigned to project activities and the accuracy of their baseline values is of importance to out-turn cost predictability.

3.8.25   Cost coding at high density

3.8.25.1   By the time the work is about to start on site, all resources must be established and the resources to be employed allocated to the appropriate activities. Cost-coding structures at this level are more often assigned to resources and the accuracy of their baseline values and progress cost updates is vital to competent project cost control. It is, of course, paramount that the accuracy of the progress and cost data entered into the schedule periodically is based upon contemporary evidence and is reflective of the resources actually employed.

3.8.26   Logic

3.8.26.1   Unless there is a good reason to the contrary, in which case the reasons must be fully set out in the planning method statement, apart from the commencement milestone, every task will have a logical predecessor (e.g. access to zone B will not be available before such-and-such a date). Likewise, apart from the completion milestone, every task should have a logical successor.

3.8.26.2   Logic can be categorised in four ways:

- engineering logic
- preferential logic
- resource logic
- zonal logic

3.8.26.3   Some scheduling software products provide the facility for designating logic in the alternative, via an 'or' gate.

3.8.27 Engineering logic

3.8.27.1 Engineering logic, sometimes called 'hard logic', is that which is indisputably essential to the process (e.g. foundations must precede superstructure, superstructure must precede fitting-out and so on). This logic is immutable.

3.8.28 Preferential logic

3.8.28.1 Preferential logic, sometimes called 'soft logic', is that which relates to managerial processes rather than engineering. For example, 'part A' of land clearance will be planned to precede 'part B' of land clearance, and 'part A' of the foundations will be planned to precede 'part B' of the foundations. Provided that it is done in good time so as to avoid unproductive resourcing, there is no practical inhibition to changing preferential logic to, in the same example, 'part B' preceding 'part A'. If there is a practical inhibition to such inversion, the logic is not preferential logic.

3.8.29 Resource logic

3.8.29.1 This is a variety of preferential logic whereby, for managerial reasons, particular resources are to proceed in a particular sequence. For example, a particular resource will commence in 'area A' and will be planned to proceed to 'area B'. Provided that it is done in good time so as to avoid unproductive resourcing or any other loss of efficiency, there is no practical inhibition to changing resource logic to, in the same example, the resource commencing in 'area B' and proceeding to 'area A'). If there is a practical inhibition to such inversion, the logic is not resource logic.

3.8.30 Zonal logic

3.8.30.1 This is a variety of preferential logic in which, for managerial reasons, one part of the physical work is required to precede another. For example, work in the location of 'building A' will be planned to precede commencement of work in 'building B'. Provided that it is done in good time so as to avoid unproductive resourcing, there is no practical inhibition to changing zonal logic so that, in the same example, 'building B' is carried out before 'building A'. If there is a practical inhibition to such inversion, the logic is not zonal logic.

3.8.31 Logical possibilities

3.8.31.1 Whichever logic is adopted, the sequence illustrated must mirror the planned intent for the future conduct of the work from design through procurement and to work on site. The various possibilities are as follows.

3.8.32 Start-to-start

3.8.32.1 In the relationship shown in Figure 19, below, Activity B can start at the same time as Activity A but not before it.

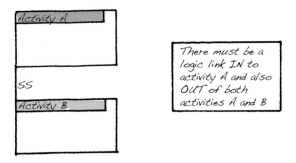

Figure 19 Start-to-start relationship.

3.8.33 **Finish-to-finish**

3.8.33.1   In the example of a finish-to-finish relationship in Figure 20, Activity B cannot finish until Activity A has finished. It implies that B can finish at the same time as A, but not before it.

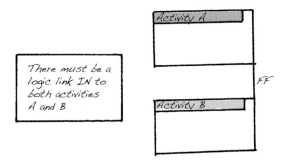

Figure 20 Finish-to-finish relationship.

3.8.34 **Finish-to-start**

3.8.34.1   The convention in Figure 21 shows the normal sequential relationship of one activity following another. For example, Activity B cannot start until Activity A has finished.

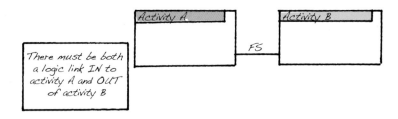

Figure 21 Finish-to-start relationship.

3.8.35 **Start-to-finish**

3.8.35.1   The convention in Figure 22 shows the unusual sequential relationship of one activity unable to finish until after the preceding activity has started. For example, Activity B cannot finish until after Activity A has started. It will be a rare occasion on which this represents a reasonable logical progression.

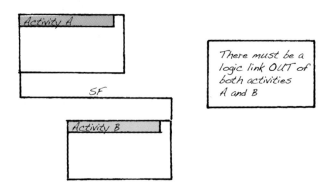

Figure 22 Start-to-finish relationship.

3.8.36 **Lags**

3.8.36.1 A lag is not an activity, nor is it a substitute for an activity.

3.8.36.2 Time lags are used in scheduling to indicate a duration following the start, or finish of a predecessor at which a successor may logically start (sometimes called a 'lead'), or a duration following the start or finish of a predecessor at which a successor may logically finish.

3.8.36.3 In reality, the time lag represents an assessment of the time necessary to accomplish the necessary predecessor section of work, the quantity and identity of which may not necessarily be specified in the schedule. If it is not specified in the schedule then the logical premises on which the lag is calculated must be identified in the planning method statement. The various possibilities include the following.

3.8.37 **Lagged finish-to-finish**

3.8.37.1 In Figure 23, 'd' indicates a finish-to-finish relationship but with a delay, that is, Activity B cannot finish until 'd' days (or whatever time period has been used) after Activity A has finished. This convention provides one of the means of overlapping the timing of activities. Instead of by reference to a time lapse, some scheduling-software products describe this as a relationship whereby Activity B cannot finish until after a proportion of Activity A has been completed.

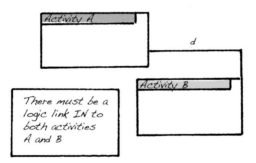

Figure 23 Lagged finish-to-finish.

3.8.38 **Lagged finish-to-start**

3.8.38.1 In Figure 24, 'd' indicates a finish-to-start relationship between Activities A and B in which B cannot start until 'd' days after Activity A has finished. An example of this might be the curing time of concrete between completion of the pour and the commencement of work on it.

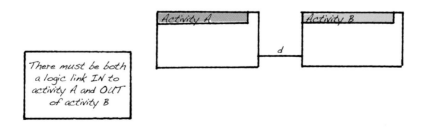

Figure 24 Lagged finish-to-start relationship.

3.8.39 **Lagged start-to-start**

3.8.39.1   In Figure 25, below, however, 'd' indicates a start-to-start relationship with the delay imposed showing that Activity B cannot start until the period 'd' has elapsed after Activity A has started. This convention provides one of the means by which the execution of activities can be overlapped. Instead of by reference to a time lapse, some scheduling-software products describe this as a relationship whereby Activity B can start after a proportion of Activity A has started.

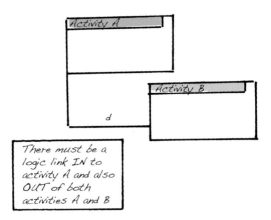

Figure 25 Lagged start-to-start relationship.

3.8.40 **Lagged start-to-start and finish-to-finish**

3.8.40.1   There may be occasions where a lag is required on both the start and the finish of related activities. This is achieved by the convention shown in Figure 26, below.

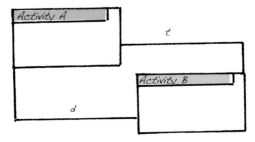

Figure 26 Lagged start-to-start coupled with lagged finish-to-finish.

3.8.40.2   In this arrangement, Activity B cannot start until 'd' days after Activity A has started, and Activity B cannot finish until 't' days after Activity A has finished. For example, in the case of a pipeline, the activity 'lay pipes' cannot start until 'x' days after the start, or until 'y' days after the finish of activity 'excavate'. Thus, 'lay pipes' has a lag start of 'x' days and a lag finish of 'y' days after the start and finish date of the activity 'excavate'. Some scheduling-software products describe this as a relationship whereby work cannot start or finish until after a proportion of the predecessor or successor has started or finished, instead of by reference to a time lapse.

3.8.40.3   A chain of three or more activities, with start-to-start and finish-to-finish driving lags, is called a ladder. Because of the propensity of many scheduling-software products to give priority to driving finish-related logic over durations, if the duration of an activity in the ladder should change, these configurations can produce ridiculous results. Accordingly, ladders should be avoided unless the scheduling software being used is capable of producing a sensible resolution of the effect of logical inconsistencies[7].

3.8.40.4   Except where lags are used to represent a curing period following a concrete pour, or similar pauses in sequence, as a general rule, the duration of lags should not exceed 50% of the duration of the shortest duration of the activities to which they are linked.

3.8.40.5   Some scheduling-software products attribute the calendar of the predecessor activity to the lag, some attribute the calendar of the successor and others permit an entirely different calendar to be allocated to the lag from that attributed to the predecessor and/or successor activities. Accordingly, in setting out the logical premises of the lag, its calendar should be identified.

3.8.41   Negative lag

3.8.41.1   Negative lag is the relationship between a start and finish of linked activities in which the successor cannot start until a period before the predecessor finishes. It is a logic which is impossible to perform and, although not prevented by most scheduling software, should never be used because it can falsify criticality.

3.8.42   Lags at low density

3.8.42.1   The use of lags is a useful scheduling technique at low density. Low-density schedules may legitimately contain a high number of lagged sequences, simply because at this density it is assumed that many of the activities generated will be speculative.

3.8.43   Lags at medium density

3.8.43.1   For medium-density schedules, durations will be shorter and the sequences will contain a higher number of finish-to-start sequences, but may legitimately contain some lagged sequences with activities of long duration.

3.8.43.2   Lags can be a useful scheduling technique at medium-density levels in that they tend to reduce the number of activities it is necessary to illustrate. However, at medium density there should be less of a need to maintain lags previously inserted at low level, or to introduce new lags.

3.8.44   Lags at high density

3.8.44.1   Without an accurate model of the planned intent it will not be possible to forecast when activities are due to start or finish, or to identify, from time to time, whether any activity is critical to completion. It is thus of paramount importance that the logic applied to high-density schedules accurately illustrates what is required and in high-density schedules lagged sequences will rarely be acceptable.

3.8.44.2   Unless there is a good reason to the contrary (in which case the reasons must be fully set out in the planning method statement), lags should not be used at high-density levels. At this level, all activities should be identified sufficiently clearly for the resources following the schedule to understand at which point in the process each activity must be started and finished in relation to its predecessor.

---

[7]   One way of achieving this is to set the float calculation to 'most critical' where the software permits it.

3.8.45 Constraints

3.8.45.1   In most scheduling-software products there are options available for manipulating the effect of logic with date-related, or float-related constraints. The definitions of constraints and the way they are treated will differ between scheduling-software packages. However, in all cases the effect of a constraint is to override what would otherwise be the logic of the network.

3.8.45.2   The effect of applying a constraint to an activity is to inhibit the activity from obeying any logic inconsistent with that constraint. The consequence of that may be that the schedule will be unable to predict, according to the logic, the dates upon which activities are due to start and finish and it may also give a false impression of criticality. Accordingly, optional constraints, if used at all, must be used with great care and be justified in the planning method statement.

3.8.45.3   The date for the commencement or finish of an activity, or chain of activities may need to be established as a constraint when resources are expected to be unavailable until a specified point in time (referred to as a start-on or start-after constraint), or where events are to be scheduled at their latest possible dates (referred to as an as-late-as-possible constraint).

3.8.45.4   Constraints can usefully be categorised as:

- flexible, in which activity start and finish dates will change according to any changes in logic and the associated resources;

- moderate, in which the activity start and finish dates will respect some changes in logic and the associated resources, but not all;

- inflexible, in which the activity start and finish dates are dictated solely by the constraint and will not change to reflect changes in logic and the associated resources.

3.8.46 Flexible constraints

- **As-soon-as-possible** – This schedules the earliest possible start and finish dates for the activity. The early start and early finish are set equal to the late start and late finish and float is eliminated from the activity and all its predecessors. In some software products this is the default constraint by which the critical path is calculated in a forward pass.

- **As-late-as-possible** – This schedules the latest possible start and finish dates for the activity. The early start and early finish are set equal to the late start and late finish and float is eliminated from the activity and all its successors. In some software products this is the default constraint by which the critical path is calculated in a backward pass.

3.8.47 Moderate constraints

3.8.47.1   These should generally be avoided unless essential. They are:

- **Zero-free-float** – This constraint will schedule the activity so that it finishes immediately prior to the commencement of the successor activity. This is an acceptable way of identifying, for example, the planned date of information release preceding a dependent activity (often coupled with a defined lag representing a mobilisation period), or for modelling 'just in time' materials' deliveries.

- **Finish-no-earlier-than** – This indicates the earliest possible date on which the activity can be completed, and the activity cannot finish at any time before the specified date. The occasions on which this is an appropriate constraint to apply will be rare.

■ **Finish-no-later-than** – This indicates the latest possible date on which the activity can be completed, but the activity can be finished on or before the specified date. The occasions on which this is an appropriate constraint to apply will be rare.

■ **Start-no-earlier-than** – This indicates the earliest possible date on which the activity can begin, and the activity cannot start at any time before the specified date. This is often used for the commencement of different phases of operation within a schedule where the phases otherwise have no logical dependencies.

■ **Start-no-later-than** – This indicates the latest possible date on which the activity can begin, but the activity can start on or before the specified date.

3.8.48 Inflexible constraints

3.8.48.1 These inhibit the schedule from reacting to change and hence must never be used. They are:

■ **Must-finish-on** – This constraint indicates the date on which the activity must finish. This takes priority over other scheduling parameters such as activity dependencies, lead or lag time, and resource levelling.

■ **Must-start-on** – This constraint indicates the date on which the activity must begin. This takes priority over other scheduling parameters such as activity dependencies, lead or lag time, and resource levelling.

■ **Zero-total-float** – The effect of this is to set the early and late dates to be the same and to render critical the activity to which it is applied, together with its controlling predecessors and successors.

■ **Expected-finish** – This constraint identifies a finish date in the future, that is, to the right of the data date. Its effect is to change the duration of the activity to which it is applied to span between its early and expected finish date.

■ **Mandatory-project-finish** – Some scheduling-software products permit finishing constraints to be assigned not only to activities but also to the project as a whole. The effect of such a constraint is to determine the latest permissible finish date of the project. This is usually set up in a way that is undetectable except by electronically interrogating the schedule set-up.

3.8.49 Inflexible combinations of constraints

3.8.49.1 These also result in inhibiting the schedule from reacting to change and hence must not be used. They are:

■ start-no-earlier-than in combination with start-no-later-than on the same date; this has the same effect as zero-total-float.

■ finish-no-earlier-than in combination with finish-no-later-than on the same date; this has the same effect as zero-total-float.

3.8.50 Float

3.8.50.1 Float occurs in a critical-path network as a result of the calculations made by reference to activity durations and logical sequence. Designated non-working periods (such as those identified as religious, industrial or statutory holidays, or weekends) are not float and should not be treated as such.

3 Developing the time-model

3.8.50.2   Consistent with the type and quantity of float, activities that are in float will be able to absorb a degree of flexibility in their start and finish dates without affecting the critical path. However, because the available float at any one time will not remain constant and it is not normally preserved by the construction contract for any one party's use at any time, the availability of float on some, or even on all, activities in a network should not be seen as a substitute for contingency planning.

3.8.50.3   The degree of float, which would otherwise exist, will be reduced as a result of resource levelling or the introduction of contingencies.

3.8.50.4   There are three types of float that are important in critical-path-networks scheduling. Their relationship to activities and to completion are illustrated in Figure 27.

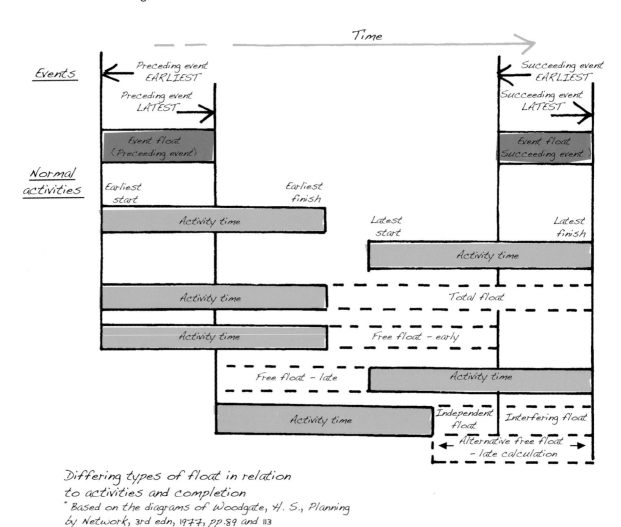

Differing types of float in relation
to activities and completion
" Based on the diagrams of Woodgate, H. S., Planning
by Network, 3rd edn, 1977, pp.89 and 113

Figure 27 Differing types of float in relation to activities and completion.

3.8.51   **Free float**

3.8.51.1   Free float is the period by which an activity can shift in its timing without affecting any other activities.

3.8.52   **Total float**

3.8.52.1   Total float is the period by which an activity can shift in its timing without affecting the relevant completion date.

3.8.53 **Negative float**

3.8.53.1   Negative float is an indication that an activity is logically scheduled to occur at a date later than an imposed date constraint will permit.

3.8.54 **Risk and contingencies**

3.8.54.1   Contingency planning is the development of a strategy to minimise the effects of intervening events which could possibly occur to interfere with the smooth running of the project at some time between inception and completion. A contingency is a planned allotment of time, which may be taken up to accommodate the occurrence of an intervening event.

3.8.54.2   Accordingly, in the same manner that cost budgets usually have an allocation of funding called 'a contingency sum', which the employer may rely upon to spend against unforeseen additional work, the schedule must have strategically placed contingency activities to absorb the time effect of intervening events which are at the employer's risk under the contract.

3.8.54.3   A prudent contractor will also make allowances for the risks it bears in the management and distribution of the resources, their variable productivity in differing circumstances and the quality of the work it carries out.

3.8.54.4   Only that party who is contractually liable for the consequences of the risk maturing[8] can properly determine the quantity and distribution of the contingency it perceives to be required from time to time. Accordingly, contracts should (and generally do) make clear who is contractually liable for the consequences of the risk maturing and thus who owns any contingency provided for the risk.

3.8.54.5   Some scheduling software will offer a choice as to whether activities are, by default, to be scheduled as early as possible, or as late as possible. Others default to one or the other. Where an activity would otherwise be scheduled as late as possible, the introduction of a contingency period buffering its end date will have the effect of scheduling the planned commencement of the activity earlier than would otherwise be the case. The effect of this will provide for a degree of delay in the completion of the activity to be absorbed by the contingency period.

3.8.54.6   Designated non-working periods such as industry-related or statutory holidays, or weekends, are not contingency periods and should not be treated as such.

3.8.54.7   Contingency periods should be designed to be identified separately for both the employer's and the contractor's risks and for those risks that are related to:

- an activity or chain of activities;
- contractor, subcontractor, supplier or other resource;
- an access or egress date, or date of possession, or relinquishment of possession;
- the works, any defined section, and any part of the works.

3.8.54.8   Typically, those general risks for which it may be appropriate to allocate a time contingency for the likelihood of time being taken up on one or more chains of activities are those listed in Appendix 1. Reference should be made to the contract in question as to which party is to bear which risks. Special risks which derive

---

[8] In an ideal world, this is also the party most able to manage it.

from the particular contract type and conditions of operation may also require contingency periods to be provided for.

3.8.54.9   Contingency periods should also be provided to cater for a delay in the completion of:

■   the work as a whole;

■   any sectional completion date;

■   any key date; and

■   the work of any contractor, subcontractor, or supplier.

3.8.55   Contingencies at low density

3.8.55.1   Much has been written in guides to project management upon the subject of the quantification of risk. It is not the intention of this publication to reproduce that but further reference could be made to:

■   Project Management Institute Book of Knowledge

■   Association of Project Management Guide to Risk Management

3.8.55.2   At the lowest level of density contingencies are likely to be the longest in order to provide some accommodation for the unknown aspects of the schedule. Because of the absence of precision at this level of density, the separately allocated contingencies to one party or the other may both be arrived at by a formula adjustment.

3.8.55.3   One way of identifying contingencies at this level is to use a formulaic approach such as Monte Carlo analysis to allocate an additional period to the presumed activities. The Monte Carlo algorithm randomly generates values for uncertain variables over and over again to generate model contingency periods.

3.8.55.4   In order for a Monte Carlo simulation to arrive at the appropriate contingency margin, a range of dates or durations must be assigned to each scheduled activity. From within those ranges, for each iteration, the given mathematical model will then randomly select a value for the duration of each activity.

3.8.55.5   There are many ways in which the range of values may be attributed to an activity for each iteration ranging from linear to parabolic progression. However, a starting assumption for each project activity may often be a 'triangular distribution'; the distribution is described as triangular because that is the shape of the probability graph for the task duration.

3.8.55.6   For example, in Figure 28 the graph shows the triangular distribution for a task with a minimum duration of 8, most likely of 14 and maximum of 24.

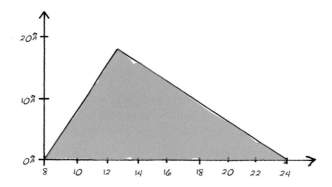

Figure 28 Triangular distribution of duration risk.

3.8.55.7   From Figure 28 it can be observed that there is approximately an 18% chance of the duration being 14 days, approximately an 11% chance of the duration being 17 days and there is a 0% chance of the duration being either less than 8 or more than 24 days' duration.

3.8.55.8   A description of the above distribution may be expressed as: 'the task is likely to have a duration of 14 days. There is nil chance of it taking more than 24 days, and nil chance of it taking fewer than 8 days. The task will probably be somewhere between 12 and 16 days in duration'.

3.8.55.9   In order to apply a contingency to that, the activity duration will thus be 16 days, or a task of 14 days' duration and a contingency period of 2 days following it. Because this is calculated formulaically rather than by reference to specific risks, it must be considered at this level to be a notional contingency to accommodate all risks of both parties.

3.8.56   Contingencies at medium density

3.8.56.1   There is little scope at this level for notional formulaic calculations to accommodate unknown and unquantifiable risks.

3.8.56.2   At this density contingencies must be clearly allocated to one party or another. There must be no contingency that is unallocated to an owner.

3.8.56.3   At medium density the risks should be clearly identified and a rational explanation set down in the planning method statement of the manner in which the possibility of the risk maturing has been allowed.

3.8.56.4   Allowances should appear at this level for such unknowns as prime cost and provisional sums (which are not required to be measured by the contractor, according to the definition under the relevant standard method of measurement, if any), approximate quantities, and other allowances for time to be taken up in the expenditure of other financial contingency sums.

3.8.57   Contingencies at high density

3.8.57.1   Simply because the predictive period is brief, the risks that need to be accounted for at high density will tend to be significantly fewer than at other densities.

3.8.57.2   At this density contingencies must be clearly allocated to one party or another. There must be no contingency that is unallocated to an owner and none that are not clearly justified in the planning method statement.

3.8.57.3   There may legitimately be risks such as adverse weather, unforeseeable ground conditions, utilities and third-party projects, plant breakdown, rework or absenteeism that may need to be allowed for at this density, but there should not be the need for design risk contingencies or implied variations at this stage.

3.8.58   The critical path

3.8.58.1   The critical path is the longest sequence of activities from commencement to completion of a key date, section, or completion of the works as a whole. In relation to each, it is that sequence of activities which will take the longest to complete or, put another way, the sequence of activities which will determine the earliest possible finish date. Hence, it is timely commencement and completion of those activities on that path which will secure completion of the key date, section, or the works as a whole on time.

3.8.58.2   As activity start and/or finish dates change, or intervening events affect the sequence of work, the critical path will change to add some activities that were hitherto not critical, and to remove from the critical path other activities which previously were critical. Hence, an identification of the critical path and understanding of what calculations and settings went into the making of that identification from time to time are very important first steps in the time-management process.

3.8.58.3   In respect of each path, the scheduling software calculates the scheduled dates of activities by taking into account the durations of all activities and the logic between them. It is a two-process calculation: the first (or forward) pass starts from the beginning of the schedule and continues to the end, calculating the earliest start and finish dates of each activity according to the logic; then the second (or backward) pass starts from the end of the schedule and regresses to the start, calculating the latest start and finish dates.

3.8.58.4   When the scheduling software determines that the earliest and latest dates are the same for any activity start or finish, it indicates them and the logic between them as being critical and identifies that the start, finish, or the whole of the activity must be completed on time, in order to complete the work in the shortest time.

3.8.58.5   When the latest start and finish dates are later than the earliest start and finish dates, the scheduling software calculates the difference and identifies it as free float. This float can then be absorbed by slippage during the course of the works, until the earliest and latest dates are the same, at which time the start and/or finish of the activity will become critical.

3.8.58.6   In the event that a completion date is fixed by a mandatory constraint (or combination of otherwise flexible constraints, which have the same effect) and the earliest start or finish dates of an activity are later than those fixed by the constraint, the scheduling software shows those activities to have less than zero total float, indicating that delay to completion is then predicted. At this stage, depending upon how many activities remain to be completed, a higher incidence of criticality may be evident, but not always.

3.8.58.7   Because different scheduling-software products have different scheduling algorithms and, amongst other things, deal in different ways with conflicts, constraints and calendars, different software will often predict a different critical path from the same data. Accordingly, identifying the critical path requires some understanding of the methods and algorithms used by the software in producing the results of its calculations.

3.8.58.8   Float is diminished by the effects of resource levelling and this will be likely to affect the critical-path calculations. Total float values on a critical path may not always be the same because of the effects of resource constraints, calendars and scheduling method adopted. Accordingly, defining criticality by reference to float values alone is not recommended.

3.8.59   Planning method statement

3.8.59.1   The planning method statement is the written description of the time management strategy, the planning logistics and the scheduling assumptions for the various parts of the works. The planning method statement is used to facilitate control of the operations and to ensure that all concerned have a clear understanding of why the work is planned and scheduled in the way it is.

3.8.59.2   Once the risk assessment of proposed operations has been carried out, the planning method statement will set out the reasoning behind the approach to the various phases of construction and list the work encapsulated in the schedule activities forming the basis of activity logic.

3.8.59.3   The longest path to each key date, section, phase or project completion date should be described in a short paragraph, and a summarised version illustrated, either as an extract from the scheduling software or pictorially.

3.8.59.4   The planning method statement must be reviewed, revised and updated, together with the schedule. The detail of the planning method statement should be consistent with the density of the schedule.

3.8.60   Method statement at low density

3.8.60.1   At this level, the planning method statement can be expected to contain:

- a description of the work to be carried out, including design, procurement and development strategy and constraints;
- third-party and neighbour interests and interfaces;
- description of the approach to risk assessment and the risks identified (e.g. work affecting river crossings in the fish-breeding season, earthworks in winter and weather cycles to which any resource is sensitive);
- an assessment of contingencies to be allowed for those risks;
- a description of the activities contained in the schedule by reference to their activity ID codes;
- the work-breakdown structure;
- calendars for working weeks and holiday periods;
- generic resources anticipated and anticipated resource constraints;
- permits and licences required and the decision periods expected in relation to each application and their dependencies;
- material and equipment restrictions and availability;
- the approach to utilities and third-party projects, licences and restrictions such as power outages;
- the approach to schedule review, revision and updating;
- activity codes applied;
- cost codes applied;
- details of the phasing and zonal relationships of the project;
- principal methods of construction;
- details of major plant requirements;
- site management, logistic assumptions and site welfare, temporary works including scaffolding, access and traffic management;
- health and safety;
- environmental considerations;
- principal methods of procurement and their effects;

■ the methods used to estimate durations;

■ the assumed sequencing logic and an explanation of any logical constraints;

■ description of the critical and near-critical paths to key dates, sectional completion dates and completion of the works as a whole; and

■ reporting formats, communications strategy and information format.

3.8.61 Method statement at medium density

3.8.61.1   At this level, detail will be added to the outline information given in the low-density schedule, including any additions, deletions, amendments or refinement of that information. It should also include:

■ identified specialist contractors, subcontractors and suppliers;

■ key-trade-interface management strategy;

■ design-and-procurement-interface management strategy;

■ limited possessions;

■ planned overtime;

■ temporary traffic diversions and plant maintenance down time;

■ resources anticipated and any anticipated resource constraints;

■ material and equipment restrictions and availability;

■ utilities and third-party projects, licences and restrictions such as power outages;

■ schedule review, revision and updating;

■ the methods used to estimate durations; and

■ details of plant requirements and their assumed productivity, down time and maintenance.

3.8.62 Method statement at high density

3.8.62.1   At this level, the planning method statement will be refined to detail the activities to be carried out in the short term, including any additions, deletions, amendments or refinement of the medium-density planning method statement. It will also include a definition of:

■ resources to be employed;

■ productivity quotients expected;

■ detailed calculations of activity duration;

■ detailed methods of construction; and

■ details of plant requirements and their productivity, down time and maintenance.

3.8.63 Quality assurance

3.8.63.1   Because competent contract administration, timely completion and financial success of the project ultimately depend upon the integrity of the schedule, it is essential that the planning strategy and schedule are reasonable, rational and achievable.

3.8.63.2   Quality assurance audits are best performed by an independent party, unassociated with the project or any of its participants; the absence of any implied knowledge helps to ensure that the right questions are asked and appropriate and understandable answers are given.

3.8.63.3   The first validation should be carried out at inception.

3.8.63.4   The assurance of quality and integrity in the schedule is achieved by both an initial validation of the development schedule and continuing checks against review and revision schedule updates, which audit the contemporaneous database for accuracy and completeness against work content and actual performance.

3.8.63.5   Subsequent and more detailed audits should then be carried out on the working schedule prior to any work stage commencing.

3.8.63.6   It should be borne in mind that the longer the period between audits, the longer the examination will take and the more serious are likely to be the consequences of any errors found. Accordingly, detailed audits should be carried out against revisions and updates at intervals no greater than two to three times the reporting period, depending upon the nature of the work being carried out.

3.8.63.7   Typically, the scope of validation will include examination of:

- buildability

- schedule content

- schedule integrity

3.8.64   Review for buildability

3.8.64.1   This review typically encompasses an assessment of the appropriateness and the degree to which the schedule has addressed the planning method statement and, in particular, the following:

- employer's requirements

- design and integration of specialist design

- quality specifications

- procurement

- construction

- health and safety

- environmental matters

3.8.64.2   The purpose of the buildability review is to minimise the possibility of intervening events occurring as a result of errors or omissions, and to assess the reasonableness of contingencies for the likelihood of delay occurring as a result of errors and omissions not being identified before work on site commences.

3.8.64.3   Experience in the discipline and project type to be assessed is vital to the success of this task, since it is as a result of experience that potential areas of difficulty can be identified prior to construction and the causes of failure of previous projects of like kind can be avoided. Accordingly, a successful buildability review will involve consideration by a wide variety of construction professionals including, in relation to any defined area for consideration, those whose speciality is:

- design

- specification

- production information

- procurement

- cost estimating

- scheduling

- information management

- quality assurance

- health and safety

- environmental matters

- sustainability

- energy

3.8.65 Review for schedule content

3.8.65.1   The schedule must illustrate a realistic and practical project plan showing how the project is intended to be, in a form that is sufficiently accurate for its identified density.

3.8.65.2   It will include a review of the specific activities, reasonable durations for the activities, and the planned sequence of work for the project. The logic or sequence of work activities should represent how the project is intended to be built and how the various activities are interrelated. The primary objective of a schedule content review is to determine that the project schedule contains the information necessary to render it fit for its intended purpose and is accurate, logical and achievable.

3.8.65.3   The purpose of this review is to establish that, for any given density, the planning method statement and schedule adequately describe what is intended as a time-model. The review will include examination of:

- planning strategy, means, methods and assumptions;

- work-breakdown structure;

- coding structure;

- estimates and calculations of resources, productivity and durations;

- engineering logic, resource logic and preferential logic;

- contractor, subcontractor, work-package logic, production and space restraints;

- constraints: milestones and performance duration;

- cost coding, budgets and earnings;

- calendars;

- design fabrication, procurement and delivery lead times;

- submittal and approvals' schedule;

- risk register and contingencies; and

- the critical path to each key date, section, or completion date of the works as a whole.

3.8.65.4   Where as-built data is incorporated into an update, the review will also include validation of:

- submittal and approvals' register;

- as-built and progress data;

- performance and productivity data; and

- cost budgets and earnings.

3.8.65.5   Where intervening events have occurred, the review will also include validation of:

- event register;

- event synopses and fragnets;

- impact methodology and calculated effect;

- the critical path to each key date, section, or completion date of the works as a whole;

- recovery and acceleration strategy and proposals; and

- recovery and acceleration monitoring.

3.8.66   Review for schedule integrity

3.8.66.1   The purpose of this review is to establish that the schedule and planning method statement are completed to a density appropriate for their use and that the schedule will respond dynamically to change.

3.8.66.2   Many scheduling-software products contain facilities for distorting the schedule or for concealing its inadequacy. In inexperienced hands this can result in schedules which conceal deficiencies in logic, durations, progress or content, which may not be uncovered until it is too late to make the necessary corrections.

3.8.66.3   Accordingly, the purpose of a review of schedule integrity is to ensure that the schedule used as the time-model can function physically as such and can be used reliably and safely, at any time, to predict the consequences of actions and inactions.

3.8.66.4   The integrity of the schedule is of paramount importance in predicting consequences because it is the skeleton for the calculation of the effect of any intervening events added to it and the baseline from which efficacy of corrective action can be estimated. If the working schedule does not react dynamically to change, or does not react logically, the calculation it produces will be of little help in identifying cause and effect, or predicting the future conduct of the work.

3.8.66.5   Accordingly, the process of reviewing schedule integrity is to interrogate the schedule to identify any faults which might inhibit its use as a time-model. In principle the review will include investigation of:

- constraints and constraint types

- logic including:

  ○   open ends

  ○   long lags

  ○   negative lags

  ○   ladders

  ○   scheduling options

  ○   critical paths

3.8.67 **Review for constraints**

3.8.67.1   Any chain of activities that is identified as critical but does not start at the beginning and finish at the end of a schedule is likely to be illustrated as critical only because of a manually applied constraint which has dictated that effect.

3.8.67.2   Because such constraints distort float calculation and hence, criticality, all inflexible constraints and combinations of constraints which produce inflexibility should be removed.

3.8.67.3   All other manually applied constraints should be removed and replaced with logic links wherever possible.

3.8.67.4   Where constraints are applied to a contractual milestone date for such things as employer-provided items, or approvals, they should not be logically connected to the network. On the other hand a duplicated milestone with no constraints should be logically connected to the activities for which such items are required. On such occasions, 'start-no-earlier-than' constraints are reasonable constraints to use for the contractual date. For the milestone logically connected to the network, a zero free float, coupled with an appropriate lag, if required, is useful in permitting a milestone (for example, an information release date) to move dynamically with the timing of the activity to which the predecessor logically relates.

3.8.68 **Review for open ends**

3.8.68.1   Open ends are sometimes referred to as dangles. Activities that have either no predecessors to their start (leading open ends), or no successors from their finish (trailing open ends), must be identified and corrected.

3.8.68.2   Because many scheduling-software products identify as open ended only those activities with no predecessors, or successors, rather than those activities with no predecessor to the start, or no successor to the finish of the activity, review of the free-float values against each activity should be made in order to ascertain the likelihood of incomplete logic. (For example, an activity with only a start-to-start link may not be indicated by the scheduling software to be an open-ended activity, notwithstanding that it has no closing logic for its finish.)

3.8.68.3   Where the activity concerned is the starting milestone identifying commencement of the project, there is no logical predecessor; other leading open ends should be identified and logic added to remove them where possible.

3.8.68.4   Ultimately, completion of every activity is a precondition for final completion and all trailing open ends must be removed.

3.8.69 **Review for long lags**

3.8.69.1   Lags represent an impediment to the progress of a successor, either as a portion of a predecessor, which impedes the successor's progress, or as an unscheduled activity (such as concrete curing).

3.8.69.2   Lags are sometimes wrongly used for off-site procurement, or for activities by parties other than the schedule's author, or they may be used to represent unknown scopes of work to be detailed as the information comes to hand, but these are practices to be avoided and the logic corrected with appropriate activities.

3.8.69.3   Lags cannot be statused during schedule updates; accordingly a lag should be replaced with a complete activity wherever possible.

3.8.69.4   If any lag is longer than half the duration of the shortest activity to which it is linked, it is usually an indication of faulty logic and should be thoroughly investigated.

3.8.70 Review for negative lags

3.8.70.1   These are used to demonstrate that the timing of future activities is to determine past activities: a sequence which is impossible to perform.

3.8.70.2   Negative lags are fatal to a time-model, which depends upon its logic for the result it displays, because they will distort float calculations and hence criticality.

3.8.70.3   Negative lags should be removed and replaced with suitable logic. Typically, the predecessor activity should be split into two activities to resolve the logic into a finish-start, coupled with a start-start and finish-finish.

3.8.71 Review for ladders

3.8.71.1   A ladder is a series of three or more activities, all of which are linked start-to-start and finish-to-finish with driving or near-driving relationships. Because this overlapping type of structure is so common and useful in construction projects, particularly in schedules of low and medium density, understanding how the software 'works' in this type of situation can be critical to the validity of the time-model.

3.8.71.2   However, depending upon the software used and the way it is configured, the shortening or lengthening of the duration of an activity in a ladder may produce ridiculous results, catapulting its successors forward in time (see Figure 29, below).

3.8.71.3   The reason for this is that in case of inconsistency between driving-start logic, activity duration and driving-finish logic, many scheduling-software products dictate priority to driving-finish logic[9]. On the other hand some software products provide the alternative facility for taking activity durations as a priority by dictating them to be intermittent instead of contiguous, or have the facility for

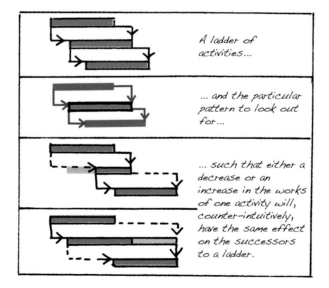

Figure 29 The problem with ladders.

[9] This can be avoided by setting float calculation to most critical, if the software provides for it.

designated 'ladder activities' such that they act in a feed-type manner so that progress in successor ladder activities is proportional to predecessor activities.

3.8.71.4   Ladders should be investigated and, where the software does not handle them adequately, they should be broken into their constituent activities in more detail.

3.8.72   Review for scheduling options

3.8.72.1   The scheduling options adopted within the scheduling software must be reviewed in order to gain an understanding of the way the schedule operates and makes its calculations.

3.8.72.2   Identify whether the activity durations are based upon interruptible or contiguous activity durations. The normal default method of calculating activity duration is contiguous.

3.8.72.3   Check whether the scheduling method is set to progress override or retained logic.

3.8.72.4   Consider whether lag is calculated from early start or actual start.

3.8.72.5   Review the method of calculating total float.

3.8.73   Review for critical paths

3.8.73.1   A review of all critical paths from commencement to completion of each key date, section, phase and the project as a whole is essential.

3.8.73.2   If an unbroken path cannot be traced from completion back to commencement, it will be because there are constraints in the schedule distorting the logic, or incomplete logic.

3.8.73.3   Missing logic should be added and, if there are constraints in the schedule, these should be removed and replaced with logic, if at all practicable to do so.

3.8.73.4   The percentage of activities on any critical path will change according to the density of the schedule. Commonly, at low density 50% of the schedule could be critical whereas in medium density around 15% of activities being critical to completion will be more usual. If there is significantly more or less than that, the reason for it should be investigated.

3.8.73.5   Ideally, all the critical paths should be reviewed and discussed in detail with the employer, design team and project-management team to gain understanding, both before and after any changes to the schedule have been undertaken.

3.8.74   Documentation of corrections

3.8.74.1   Any corrections required to facilitate the schedule's use as a time-model must be justified and documented in a corrections' log. Every required correction should be documented clearly and succinctly and, when completed, the corrective action taken should also be logged for review.

3.8.74.2   The effects of any corrections, and any revisions required as a result of those corrections, should be identified in a revision to the planning method statement.

3 Developing the time-model

4 Managing the time-model

4.1 Introduction

4.1.1   In summary, management of the time-model is the process of managing the schedule during its delivery which, in turn, involves:

- review and revision of the assumptions used to produce the schedule;

- collection of production records;

- monitoring the work in progress;

- updating the schedule;

- identifying intervening events;

- impacting them on the schedule;

- implementing recovery and/or acceleration; and

- revising the planning method statement to record what has changed and the reasons for it.

4.1.2   The employer and its professional team are entitled to know what the contractor has achieved to date, and whether it is on target to complete by the various key dates and contract completion date. The working schedule is thus not just the schedule of the contractor's intentions, but a fundamental management tool for the employer and the design team as it provides:

- the essential information regarding the timing and interface of their continuing obligations; and

- the calculating mechanism in the event that the contractor becomes entitled to an adjustment of the time to complete.

4.1.3   For its success, time-model management depends upon the quality of the working schedule and the ability to absorb and integrate the information collected during the management process. Hence, the party managing the schedule has a duty to ensure that the schedule is robust and as accurate as it can be.

4.1.4   As better information becomes available, the schedule must be revised to incorporate it. Revision is not updating, nor is it the impacting of intervening events. It is the process of adopting a change in the assumptions previously used as to the detailed planning of the contract works, as better and more precise information becomes available during the course of the project.

4.1.5   The relationship between the different levels of schedule revision, monitoring, updating and impacting in the schedule-management process is illustrated in the flow chart in Figure 30, below.

Guide to Good Practice in the Management of Time in Complex Projects, by The Chartered Institute of Building
© 2011 The Chartered Institute of Building

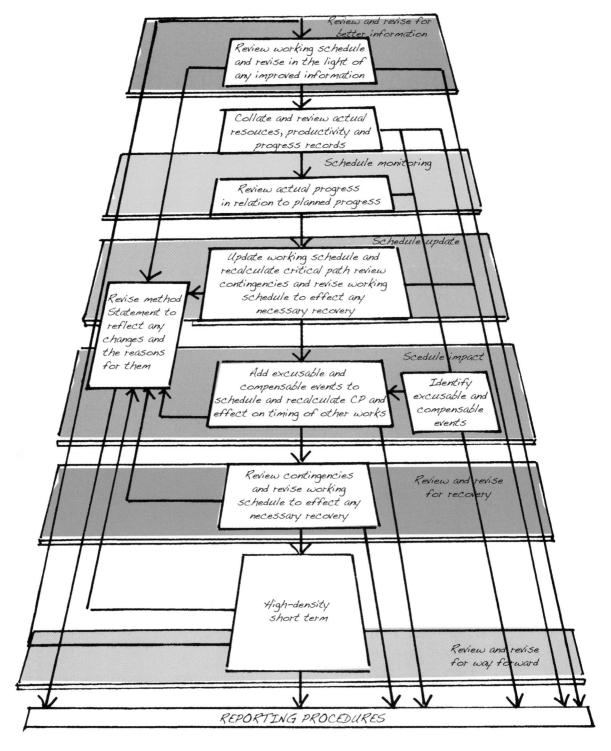

Figure 30 Relationship between review, revision, monitoring, updating and impacting.

4.1.6   The purpose of updating the schedule is to take into account progress actually achieved to ensure that the working schedule for the future accurately takes account of the effect of what has gone before and, as a result, predicts what is to be done next, when it is to be done, and what resources are to be involved. That is essential for competent resource management.

4.1.7   The time-management process must include accurate record keeping, particularly of progress. This information provides essential data for underwriting forecasts, calculating effects of change and lessons learnt. Other key aspects of good record keeping are facilitating the calculation of the effect of an intervening event:

■   progress, and the future conduct of the work;

■   on one or more key dates and the completion date; and

■   on one or more other contracts and subcontracts.

4.1.8   Provided that the schedule is up to date, it can then be used for, amongst other things, identifying the occurrence and ultimate effect of an intervening event.

4.1.9   All complex projects will be affected by intervening events from time to time. However, whilst it is normally not possible to predict when they will occur, or what they will affect once they have occurred, time cannot be managed effectively unless the consequences are calculated promptly and appropriate action taken in good time.

4.1.10   Accordingly, when relevant, the time-model can also identify and quantify (in relation to the terms of the conditions of contract) the rights of the parties to extensions of time and liquidated damages, and time-related compensation for disruption and prolongation.

4.1.11   In the circumstances of the schedule having been adversely affected by an intervening event, the time-model can also be used to manage the effects of change, for example, allocation and reallocation of contingencies, identifying the effects of alternative processes which might be adopted by way of mitigation, recovery, or acceleration, and as the model for any supplemental agreements as to change in process.

4.1.12   Finally, the as-built schedule can be used for identifying the actual sequence, resource and durations which can be used in the future to improve the predictability of the time-model.

4.2   Schedule review and revision

4.2.1   Review will not necessarily mean that a revision will follow, but experience tells us that unless a complex project is in the last few weeks before completion it is very likely that revisions to some part of the planned intent will need to be made as a result of review.

4.2.2   An effective time-model-management process will generally involve the review and revision of:

■   improved information about what is contractually intended;

■   resources and productivity as schedule density increases.

4.2.3   Managing the review and revision process

4.2.3.1   Reviewing and revising a schedule must be the result of a formalised management and governance procedure involving the project team.

4.2.3.2   Prior to commencing any review, the working schedule should be backed up so that, prior to revision, the schedule can be archived and the revised schedule can then be maintained as the new working schedule.

4.2.3.3   During revision, an audit trail should be maintained of changes, and the reasons for them, and they should be recorded in an update of the planning method statement.

4.2.3.4   On completion of all changes, the critical path should be recalculated and the resultant changes noted and recorded in the planning method statement and all parties notified accordingly.

4.2.4   General matters for review

4.2.4.1   Change in methodology

4.2.4.2   The implications of making changes in methodology should be carefully explored and considered in terms of the likely effects on time, cost, quality, risk and conflict with the contract conditions (e.g. a change from a traditional design prepared under the direction of the employer with a build-only construction contract to one of design and build is likely to affect all phases of a project).

4.2.4.3   Repetitive activities

4.2.4.4   Trends in productivity achieved, derived from the as-built productivity data, should be used to verify the planned schedule for the remainder of those activities. If a discrepancy is found between what is planned to be achieved and what, by reference to the progress records, can be proved to be achievable, changes should be made to the working schedule to accommodate the discrepancy. In relation to an activity such as piling, for example, each rig's productivity should be analysed (in order to ascertain the optimum pile-cycle achievable and the effect of any prior departures from it). The demonstrable, achieved productivity cycle can then be used to verify the activity durations planned for the remaining piling works.

4.2.4.5   If it should be found that the productivity achieved is insufficient to maintain the schedule, changes can then be made in good time to the planned resources and/or to the sequence of the works in order to bring the work back on schedule.

4.2.4.6   One of the advantages of this repetitive-sequence review is that in the case of an occurrence which disrupts productivity, the benchmark-proven optimum productivity will be the best possible baseline from which to calculate the effects of the disruptive event.

4.2.4.7   Change in activity descriptions

4.2.4.8   The scope of activity descriptions will vary depending upon the density at which the schedule is prepared. At low density the schedule will have very broad overarching descriptions, whilst at medium and high density, schedules will have very detailed, discrete activity descriptions related to the activities in a single work package. Any review and change should, therefore, take these various densities into account, ensuring that the descriptions fit the purpose for which they are intended.

4.2.4.9   Activity descriptions should be reviewed with the following questions in mind:

■   does it accurately describe the activity so that it is unambiguous in its scope and meaning, is easily understood and consistent with other descriptors?

■   does the activity or group of related activities match the planning method statement?

■   what is driving any change to the activity description?

4.2.4.10   Change in activity durations

4.2.4.11   Activity durations must be regularly reviewed at all stages of a project and updated to ensure that they are as accurate as possible, in that they reflect either improved information on an activity (e.g. subcontractor predicted resources and output data), increased or decreased content, resource availability or changed productivity quotients.

4.2.4.12 Change in logic

4.2.4.13   The logic is as important to the accurate modelling of time as are activity durations, hence any change to logic must be recorded. The review and revision of logic at any stage can improve the time line and resource usage and overcome problems of resource availability (cost, materials, plant, labour, working space and so on).

4.2.4.14   The level at which logic is reviewed and revised within a schedule will have varying degrees of impact. Changes to the logic in a low-density schedule tend to be fundamental, whereas those at medium density may affect only the relevant work package, and at high density may affect only a few activities.

4.2.4.15 Change in cost profile

4.2.4.16   Changes to schedule costing can be extensively carried out at the project-planning stage as part of the scenario-planning process in order to maximise the effective use of resources. Costing reviews and changes are likely to be closely interlinked with the overall review of the schedule, and when these take place at the working-schedule stage the effects should be carefully analysed, especially where there are contractual issues. (For example, the effect of costing the schedule on interim valuations may have to be considered.)

4.2.4.17   Review how costs are allocated, whether by unit rates, total effort, or lump sum, and where costs are allocated, whether by activity, or by resources for the activity. Check the cost centre alignment against the WBS.

4.2.5 Consequential changes

4.2.5.1 Change in floats

4.2.5.2   The results of other revisions, especially to durations, logic and resource levelling, will have an effect on float values. However, no attempt should be made to change float values by the use of constraints.

4.2.5.3 Change in critical paths

4.2.5.4   It is inevitable that the critical path will change during the life cycle of the project and therefore a review is needed in order to determine the effects of change of any part on the whole schedule and its critical path.

4.2.5.5   The critical path and its activities should also seem 'to make common sense' to an experienced reviewer and it is thus important to review the revised critical path to ensure that the sequences, logic and durations are appropriate.

4.2.5.6   It is important to be aware of the chosen software's limitations in relation to its time and/or resource-analysis calculation, and consequently the reliability of the calculated critical path indicated by the software.

4.2.5.7   As with float, the critical path should be permitted to look after itself and be a mathematical calculation of the current schedule. No attempt should be made to manipulate it by the use of constraints or defective logic such as the imposition of negative lags.

4.2.5.8   Variations between the critical path achieved as a result of review and revision and the preceding critical path will be an expression of the consequences of any changes made to the nature, duration and logic of future activities.

4.2.5.9   The critical path may consist of more than one string of logic, with other parallel activities involved, and a check should be carried out on those activities with total floats of less than (say) 10% of their durations.

4.2.5.10   Particular attention should also be given to any path which is predicted to extend beyond the contractual completion date.

4.2.5.11   It is also useful to review the driving activities to all key dates and the completion of discrete contracts and subcontracts.

4.2.5.12   In reviewing the revised critical path one should consider:

- logic (including leads and lags)
- activity durations
- float
- unexpired contingencies
- resource levels
- constraints
- method of calculation

4.2.6   Review for better information

4.2.6.1   Typically, those areas of the schedule which require attention are those concerning:

- design
- procurement
- work content
- remaining short-term work

4.2.7   Better design information

4.2.7.1   Check, review and revise, where necessary, those activities concerning:

- design methodology
- specialist input
- drawing and information control
- submittals
- approvals

4.2.8   Better procurement information

4.2.8.1   Check, review and revise, where necessary, those activities concerning:

- procurement
- works-package definition
- bill-of-quantities production
- tender appraisal
- contract and mobilisation

4.2.9   Refinements to work content

4.2.9.1   Check, review and revise, where necessary, those activities concerning:

- prime cost sums

- provisional sums

- approximate quantities

- plant availability

- construction methodology

- construction resources and productivity

- testing and commissioning

4.2.10 Review for short-term work

4.2.10.1 Check, review and revise, where necessary, those activities in the resourced short-term, high-density parts of the schedule.

4.2.10.2 Resource review (e.g. manpower, plant and materials, etc.) has an important part in ensuring that a project's objectives are met most efficiently and effectively. Resources should be reviewed in terms of:

- suitability

- type

- availability

- output

- cost

4.2.10.3 Whilst policy decisions on the type of resources intended to be used are best made at the early design stages of a project, reviewing resources at any stage can also help in producing the most efficient and time-effective schedule and is essential for the development of the short-term, high-density schedule.

4.2.10.4 In order to transition from medium to high density, close coordination is required between those who will actually carry out the work and the management team. It is essential that the work is accurately scheduled to reflect the workforce's intent and that the workforce intends to follow the work sequence with the resources allocated. Accordingly, before completing this review it is essential that the parties to carry out the work express their confidence that, in the absence of any intervening event, they can produce the resources required, achieve the productivity envisaged and work to the planned sequence with the required interfaces.

4.2.10.5 When working with resources, it is important to understand how the software analyses resource utilisation. Of particular concern should be whether, in the event of overload, the project end date will remain unchanged and resources will be increased above the indicated limit, or whether the project end date will change to reflect the time required to complete the project with the allocated resources.

4.3 Record keeping

4.3.1 Introduction

4.3.1.1 Records which cannot be retrieved are useless. Accordingly, the process of record keeping is inseparable from the process of record retrieval. It follows that in order to identify an adequate means of keeping records in the first place, consideration must be given as to how they can be retrieved and used.

4.3.1.2   The increasing speed and sophistication of databases and spreadsheets have provided the industry with the facilities for sorting and filtering data into specific reports at the press of a button and it is now no longer acceptable to proceed to keep records on paper alone. It follows that unless records are kept as database records in the first place, they must be rekeyed as database records before they can be retrieved and used.

4.3.2   Spreadsheet-recorded data

4.3.2.1   The purpose of using a spreadsheet is to facilitate simple filtering and evaluation and reporting, with the opportunity for electronic import into a database if required.

4.3.2.2   If a spreadsheet is used, the records must be kept on a standard software product and be laid out so that the information can be automatically imported into a database either at initiation, or subsequently, without subsequent rekeying.

4.3.2.3   This requires that:

- each activity identifier occupies a separate line;

- the information relating to that identifier occupies the same line as the activity to which it refers; and

- each item of information occupies a separate cell consistently from report to report.

4.3.3   Database-recorded data

4.3.3.1   Complex projects require significant record management, which can only be properly achieved by keeping the records in a database to which all those needing to see them can have access. It is the relationships in a database that facilitate the organisational and filtered view and recovery of the information (by way of the fields and values within the fields of each record), which render it the most functional of all record-keeping methods.

4.3.3.2   The advantages of keeping records in such a database framework are:

- records are stored in an accessible electronic format providing wide access for record entry and review;

- every record stored is attached to appropriate identifiers enabling organisation, filtering, targeted review and reporting;

- the record identifier and content can be controlled to avoid casual misleading record entry;

- background checking can be established to prevent record errors such as duplicate entries and excess hours for labour or plant;

- information is entered only once but can be grouped and reported to suit many output requirements;

- the records themselves can be readily monitored to ensure that they are being maintained in the manner required.

4.3.3.3   However, the overriding advantage of maintaining records on a database is that their content is drawn from one unified source and, as a result, all reports, extracts and summaries produced by the database will have mutual integrity.

4.3.3.4   Figure 31 illustrates the relationship figure of a simple database for recording project activities. This figure illustrates a database facilitating a record of data kept in three fields: 'Activity Tasks', 'When it was done', and 'What resource'.

4.3.3.5   Two fields, 'Schedule Activities' and 'What was done', serve to provide the data for 'Activity Tasks'; each of the fields called 'Activity Tasks', 'When it was done' and 'What resource' then contribute information to a table called: 'What, when and resource'.

4.3.3.6   Each of these tables is very similar to a spreadsheet. For example, when opened in the database, the table of 'What, when and resource' appears in the form illustrated in Figure 32.

4.3.3.7   The database also contains review and reporting features enabling searching and grouping of data to any number of declared criteria.

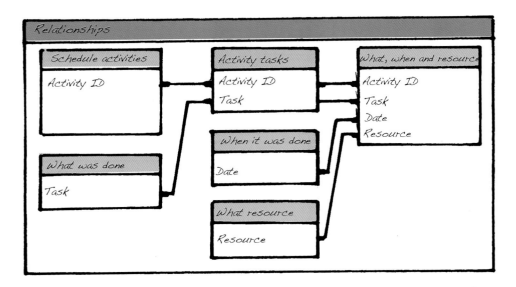

Figure 31 Simple database relationship.

What, when and resource				
Activity ID	B00170			
Activity description	Reduced level dig			
Date 17 June	Task	Resource	No. of	hours
	Cart to tip	Dumper	1	4
	Cart to tip	Machine operator	1	4
	Excavate	Excavator	1	4
	Excavate	Machine operator	1	4

Figure 32 What, when and resource table.

4.3.3.8   The database is designed to offer a user-friendly form for record entry. A simple form for making an entry is shown in Figure 33.

4.3.3.9   Whilst this simple database is primarily for the purpose of recording resources expended in relation to the scheduled activities, it is possible to develop the simple model to provide a database suitable for maintaining records of all project documentation, notices, test records, correspondence, events, issues and the like (see Figure 34).

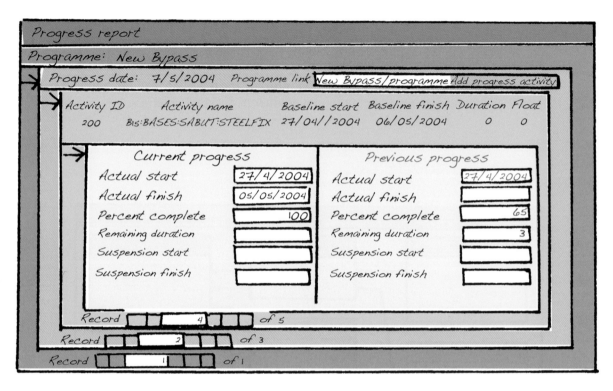

Figure 33 Simple data-input form.

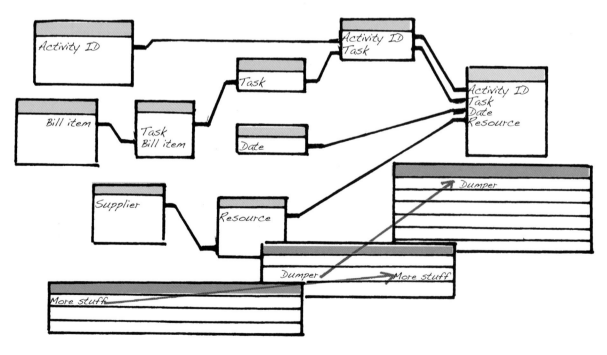

Figure 34 A database containing supplier data and bill-of-quantity data.

4.3.3.10 Where the database is network-based it can be used by others and on many projects using a wide area network. A form like this can be operated by an extranet with special access rights to a defined user group, or on a company's internet site with defined user rights.

4.3.3.11 However, networking is not the only means of interfacing with the database. This form may be linked directly to a server storing records from many sources. If email forms are used to interface with the database, an electronic-data-receipt regime will be required to transfer the emailed data into the database. For this, typically, a dedicated email address is established with automated procedures which are triggered by the incoming mail. The database can have all the integrity restraints of the network database and can automatically email replies to warn of potential errors and duplications, requesting revisions and confirmations.

4.3.4 **Record types**

4.3.4.1 The records which generally have to be kept for successful time management fall into three categories:

- progress records

- quality-control records

- information-flow records

4.3.4.2 Ideally, they should be kept on the same record-control system and inter-related through a database; otherwise, unless the separate databases are electronically integrated, double entry will be necessary to some extent, and the likelihood of error in data input will increase.

4.3.5 **Progress records**

4.3.5.1 Records of progress actually achieved will be used for identifying the start and completion dates of activities, the degree of progress achieved from time to time and for identifying the productivity actually achieved by particular resources. They can also be used for verifying productivity trends, the timing of the constituent parts of intervening events and identifying lost productivity as a result of disruption. In other words, progress records are the lifeblood of effective time management.

4.3.5.2 A consistent approach should be adopted for the gathering and collating of progress data and each data-gathering inspection is best carried out following the same inspection sequence. Because the progress records are the source from which the schedule will be updated, unless the scheduler is also responsible for collection of the data, it will be difficult, if not impossible, to support an audit trail from the fact to the updated schedule.

4.3.5.3 The optimum frequency of data collection is not related to the schedule-update period, or monitoring frequency, or the required reporting period, but will usually be related to the number of activities and the complexity of their interrelationships being carried out from time to time.

4.3.6 **Progress-record content**

4.3.6.1 The records to be kept are those that will help in the management of the works going forward and the establishment of the facts surrounding the work which has gone before. In effect, these are records concerning:

- information flow

- progress

- ○ photos

- ○ diary records

- ○ note/log books

- ○ minutes of meetings

- ○ records of transactions (receipt of information, materials, labour, plant and goods, etc.)

- ○ weather, industrial action and other issues

- ○ third-party issues

- ○ quality control

- ○ change control

- ○ productivity achieved

4.3.6.2 Unless the right information is kept in relation to the purpose for which it is to be used, no matter how accurate it is, or how well it is presented, it will be ineffective and of little use. It follows that the purpose for which the records are to be used will be the overriding factor as to their content. However, there are some types of data which are essential to any record, no matter for what purpose it is to be put. These are:

- ■ coordinating code

- ■ activity description

- ■ date of record

- ■ resource used

- ■ activity start and finish dates

- ■ the author of the record

- ■ progress data

4.3.7 Coordinating code

4.3.7.1 There must be a relationship between the planned activity (if any) and the records of work done to carry out the activity. This will ordinarily be the unique activity ID created by the project-scheduling software when the activity was planned.

4.3.7.2 If the work being recorded cannot properly be allocated to a planned activity, it must be allocated to an intervening event and identified separately.

4.3.8 Activity description

4.3.8.1 Whenever possible, the description of the work being done should follow the description of the work in the high-density working schedule. If it does not follow it, it will be because either the activities planned are in insufficient detail to be properly recorded, or the work being carried out was never planned to be carried out and constitutes an intervening event.

4.3.8.2 The date and, if relevant, the time at which the record is made must appear within the record. It seems odd to have to make the point, but unless the content of the work achieved can be related to the date it was recorded to have been achieved, nothing can be done with the record.

4.3.9 Date of record

4.3.9.1 The date of the record is an essential ingredient of a record of partially completed work: without it, the recorded proportion of work completed is meaningless. The date of a record also adds credibility to a recorded start or finish date.

4.3.10 The resource

4.3.10.1 The record must identify the quantum of resource being used to achieve the work. In so far as work in high density is scheduled according to the resources to be applied, without the identified resource actually used, the record will be ambiguous and for some purposes may be useless.

4.3.11 Start and finish dates

4.3.11.1 The date an activity started and finished must be recorded, together with the quantity of partly completed of unfinished work.

4.3.12 Author of the record

4.3.12.1 Whilst for many purposes the author of a record may be self-evident, that will not always be so and, in the event of discrepancy, the identity of the party making the record can be very important.

4.3.13 Progress data

4.3.13.1 If an activity is started and completed in its entirety in a reporting period, the work done will be self-explanatory. However, if it is not, then the quantity of work achieved in the reporting period, together with the date upon which the record was made, must be identified. Without a degree of progress identified by a certain date, for some purposes, the record will be useless.

4.3.13.2 Depending upon the facilities offered by the scheduling software, progress can usually be identified in four different ways:

■ remaining time, i.e. by reference to the expected time needed to complete;

■ expired time, i.e. by reference to the expired period in relation to the planned duration;

■ proportion of effort expended, i.e. by reference to the percentage of work done in relation to the whole; and

■ quantum of work done, i.e. by reference to the measurement of the resources used and resources remaining in relation to the measurement of work performed.

4.3.13.3 Bearing in mind that it is always the high-density part of the schedule which is to be updated, and that this is calculated by reference to the resources and their planned productivity, it is important to record the type of resources used, the quantum of resources expired against the amount of work achieved.

4.3.13.4 Whenever a repetitive cycle is scheduled, great care must be taken to ascertaining and checking the resources to be applied, the gang strength and quality control, and anticipated productivity. If it is possible to run trial cycles by way of establishing a benchmark before construction starts, so much the better. However, in practice, it is likely that, in many cases, benchmarking will have to take place over a period of time with different design teams or work gangs to identify differences between ultimate productivity and learning curves with resources of differing abilities.

4.3.13.5   In principle, apart from the basic data above, the information which must be kept will produce the answers to the following questions:

■   What and how much was done (on an activity-by-activity basis)?

■   How much duration is remaining to complete the activity (days or weeks)?

■   Who did it (the labour resource)?

■   With what was it done (the plant and material resource)?

■   When was it done (the date and timing of the activity)?

■   Where was it done (the location in which it was carried out)?

■   How was it done (the process adopted)?

4.3.14   Quality-control records

4.3.14.1   A system of records which prompts the inspection, checking and testing of the works is often a requirement of the specification and it may be required to be formally submitted as part of the employer's final acceptance procedures. Typically, such records will identify:

■   subject matter

■   specification requirements

■   test date

■   testing supervisor

■   test results

■   deficiencies identified

■   corrective action taken

■   date upon which corrective action commenced and completed

■   sign off by supervisor

4.3.14.2   As with information flow, quality-control documentation can be managed by a proprietary document management system product (DMS). However, if these records can be integrated into the same database as that used for the progress data, it will be an advantage.

4.3.15   Information-flow records

4.3.15.1   For the efficient exchange of information it is necessary to record in relation to any informational transaction:

■   unique reference number;

■   its subject matter;

■   its source;

■   what actions are needed in response;

■   who will take this action; and

■   when the action is commenced and completed.

4.3.15.2   There are many DMS products which can be used to accumulate and report upon information related to this type of data. However, if these records can be integrated into the same database as that used for the progress data, it will be an advantage.

4.4 Updating the schedule

4.4.1 Updating is not progress monitoring, nor is it schedule revision. It is simply the addition of as-built data to the working schedule and the recalculation of the critical path in the light of progress actually achieved. The records of progress are used to add as-built start and finish dates to those activities which have achieved either status, and progress data to those started but incomplete at the defined data date.

4.4.2 Updating a schedule is essential to the management of the time-model; without it, the schedule is merely a target against which historical failure might be plotted. However, by updating with progress against the progress records and recalculating the critical path, the working schedule becomes a dynamic model by which:

- predictions can be made;

- problem issues can be identified early;

- mitigating recovery and acceleration can be implemented;

- the future conduct of the works can be effectively managed.

4.4.3 The prompt identification of disruption and delay to progress is vital if the consequences are not to result in loss of quality, cost overruns and delayed completion. Once identified, difficulties can always be addressed, their potential consequences calculated, and strategies implemented to avoid or reduce those consequences.

4.4.4 An essential part of the updating process is recalculation of the critical path following the addition of progress. The recalculation of the critical path identifies the start and finish of activities to be commenced, those which are then on the critical path and provides a cogent baseline against which the effect of intervening events can be calculated.

4.4.5 The advantages of updating the schedule are that:

- the impact of change can be accurately predicted against the model, whether it be to an activity duration, sequence (logic), or resource (money, labour, plant and materials);

- resource planning is rendered more reliable because past and current experience of productivity can be better used to forecast future trends;

- the effect of changes to construction activities can be better calculated with 'what-if scenarios', enabling the project team to choose the most efficient sequences;

- potential problems are identified much earlier with resulting increased time available to mitigate any risks, deal effectively with any issues from any 'crystallised' risks and thus improve the likelihood of maintaining the project's time and resource targets; and

- the cause and effect of disruption can more readily be identified and its consequences managed.

4.4.6 In summary, the fact that high-quality-management information can be derived from the updated working schedule encourages its use by the project team for management purposes.

4.4.7 For general purposes, there should be a direct correlation between the timing of the schedule update and the reporting cycle. In other words, the data date of the updated schedule should match the reporting requirements. Legitimately, however, there may be an increase in updating frequency depending upon the construction phase and/or criticality of the work in progress.

4.4.8 During the schedule update, an audit trail should be maintained of the source of the as-built data.

4.4.9 On completion of each update, the critical path should be recalculated and recorded and the resultant changes to the critical paths noted and recorded in the planning method statement.

4.5 Change control

4.5.1 For many reasons, intervening events present a special problem in record keeping, not least of which is the difficulty of keeping abreast with changes in scope.

4.5.2 Notwithstanding that the contract may be clear and unambiguous as to who bears which risk, at a detailed level the facts may all too often be difficult to label. Accordingly, in the event of any doubt, it is prudent to maintain records of work that might legitimately fall into the class of an intervening event by attributing to it an independent activity-coding classification, which stands the work apart from that which falls under the contract. Typically, such activity coding might legitimately commence with 'EV' (as an abbreviation of 'event') where that abbreviation has not been used as part of an activity-identifier code for any part of the works.

4.5.3 The recording of change will require capture of the following information:

■ a unique event identifier;

■ description of the event;

■ originator and/or authoriser;

■ relevant contract clauses providing for extension of time;

■ relevant contract clauses providing for compensation;

■ date upon which the event is instructed/occurred;

■ responsible parties;

■ the activities added, changed or omitted;

■ the labour and plant resources for each added or changed activity;

■ the date and timing of the added or changed activities;

■ the location in which any added work was carried out; and

■ the work-flow process adopted in carrying out the change.

4.5.4 There are advantages to be gained from tracking changes that have been initiated as a separate classification from those which can be identified as potential changes.

4.5.5 As with information-flow and quality-control documentation, change-management information can be managed by a proprietary DMS software product. However, if these records can be integrated into the same database as that used for the progress data, it will be an advantage.

4.5.6   For guidance on the sort of risks often borne by the employer, reference should be made to Appendix 1. However, for information on what risks are actually borne by the employer, reference must be made to the particular contract in question.

4.5.7   Identifying intervening events

4.5.7.1   Intervening events are those occurrences which were not originally planned for. They may adversely affect productivity and/or progress and they are often difficult to identify. As a result, it is often the case that the secondary effect of many intervening events (a delay to progress) will be initially identified as a result of the schedule update. The causal event then often needs to be identified retrospectively as a result of its effect.

4.5.7.2   Intervening events can be classified in a number of ways; however, the primary classification is usually by reference to liability: those that are at the contractor's risk and those that are at the employer's risk.

4.5.7.3   However, for the purpose of management of time, it is important to consider all risks, irrespective of liability and, if disputes are to be avoided, all intervening events and their consequences should be clearly identified and agreed, on a rolling basis, irrespective of liability.

4.5.7.4   In order to undertake any meaningful analysis of the effect of an intervening event, identification of the date upon which the event itself was initiated, together with the chain of causation arising, will be of the utmost importance. This is because:

- it will clarify whether events have happened sequentially, in parallel, concurrently, or simply to keep pace with other work and assist in distinguishing the effect of one event from that of another;

- it will determine the calendar date after which an event can possibly have an effect;

- it may determine the point from which a notice under the contract may be required to be given; and

- it may determine the time at which the statutory limitation of liability provisions commence.

4.5.7.5   Whilst the detail of what is required will differ between events of different character, in principle, in relation to every intervening event, the points which must be addressed are those listed above in paragraph 4.5.3.

4.5.7.6   The logic of the intervening event should be set out clearly, together with the activity which it affects, and the way it affects it. If, for example, a planned activity has actually started when late or revised information is provided, that later information cannot logically inhibit the start of the activity. That will be so, even if the logic of the sequence was planned on the basis that all information was to be provided before the activity in question could start. Under those circumstances, if the later information has any effect at all, because the activity has already started, the effect of it will be to increase the planned duration of the activity either:

- by the time it takes to carry out the additional work in relation to the planned finish of the activity; or

- as a result of the delaying effect of the disruption caused.

4.5.7.7   Apart from changes arising out of the supply of information, there are a number of specific types of intervening events which require special consideration such as:

- voluntary and implied variations and other instructed changes;

- instructions for the expenditure of prime cost and provisional sums;

- acts or omissions of the developer, or those for whom they are responsible;

- acts or omissions of third parties;

- other occurrences;

- disruption.

4.5.8 Voluntary and implied variations and other instructed changes

4.5.8.1 Although the initiation date of an agreed variation is relatively easy to determine, many variations result from changes in design information or from instructions not expressly acknowledged as variations. Typical examples of the latter include:

- extra work arising from the unwarranted condemnation of contractually conforming work;

- the issue of amended drawings to correct a discrepancy in the employer's requirements;

- the issue of amended documentation to correct an error in the bills of quantity, or specifications;

- responses to requests for information requiring additional or changed work;

- unwarranted rejection of contractually compliant submittals.

4.5.9 Variations

4.5.9.1 Additions and omissions are both types of variations; many contracts will have detailed descriptions of what constitutes a variation and under what circumstances.

4.5.9.2 A variation cannot begin to have an effect upon the contractor's performance until the contractor knows about it and is at least in a position to act upon it. Accordingly, when variations are instructed expressly, or impliedly, the event should normally be construed to have occurred when the contractor can be shown to have received the instruction.

4.5.9.3 On the other hand, contract conditions sometimes specify when instructions are deemed to have been received. For example, the contract may specify that oral instructions are to be confirmed in writing by the contractor within a limited time before they become effective or, if not subsequently rejected within a fixed period of time, are deemed to take effect on a future date.

4.5.10 Prime cost and provisional sums

4.5.10.1 With regard to instructions for the expenditure of prime cost and provisional sums, it is good practice for the date by which an instruction for the expenditure of a provisional sum, a prime cost item, or the nomination of a subcontractor or supplier is required to be indicated in the working schedule.

4.5.10.2 Inasmuch as it requires the essential participation of both the contractor and the nominated party without which the instruction for nomination is meaningless, an instruction for the appointment of a nominated subcontractor or supplier differs in character from other instructions. The content and timing of the nominee's obligations cannot properly be ascertained until the contract administrator has stipulated their requirements in an invitation to tender, an acceptable tender has been received by the contract administrator and, as a result of the instruction to place their subcontract with the nominee, the contractor has successfully negotiated their requirements and placed the subcontract.

4.5.10.3   Accordingly, the initiation date of a delay to nomination should be taken to be the date upon which the effect of the nomination, if any, can be calculated from the terms of the acceptable nominated subcontract. In most cases, rather than the date upon which it was scheduled to be received, this can be expected to be the date upon which the instructed details of the work to be carried out are actually received to permit a proper prediction of the effect of the instruction.

4.5.11   Employer's acts or omissions

4.5.11.1   Health and safety

4.5.11.2   Where the contractor is to be excused delay to completion caused by the employer's compliance with, or non-compliance with, the CDM regulations, the employer's obligation is to ensure that the principal supervisor or the principal contractor under the regulations performs its duties competently. The initiation of the event in such circumstances will thus be the date of the breach of their obligations by the principal supervisor or principal contractor.

4.5.11.3   Late information

4.5.11.4   The date upon which a failure of the contract administrator to issue in due time any information, drawings' details or instructions requested is initiated as a developer's time-risk event, similarly depends first upon the terms of the contract. In some contracts, for example, there are two separate provisions:

- On the one hand, the employer may determine when information will be provided and set this out for the purposes of tender in an information-release schedule. In such circumstances the event will be initiated on the day after the information was scheduled to be provided in the information-release schedule, irrespective of whether it is then needed.

- On the other hand, under most forms providing for the contractor's request for information rather than the employer's offer of information, the date of initiation is usually the day after the end of the period during which the information requested should reasonably have been provided.

4.5.11.5   When it should reasonably have been provided will usually depend upon what is stated in the contract provisions and if nothing is provided, then an objectively reasonable period should be applied.

4.5.11.6   **Late response** – The failure to respond in due time to a contractor's submittal is initiated not on the date the submittal is made but on the date by which the response was due to be received by the contractor. In many contracts this is stipulated as a fixed period of time after the submittal is made for all submittals, irrespective of their content or importance. In other contracts there may be differing approval periods, depending upon the nature of the submittal.

4.5.11.7   Responses to submittals can initiate secondary processes, either as a result of the response requiring a variation in the quality or quantity of the works (in which case the principles as to when a variation is initiated apply), or as a result of the response properly requiring a resubmittal (in which case the initial submittal can be deemed to be without effect, the submittal and approval process then having to recommence from the beginning).

4.5.11.8   **Failure to grant access or possession** – Depending upon the phraseology used in the contract, a failure to give, or deferment of, possession of, access to, or egress from any part of the site is initiated on the day after it was warranted that such possession, access or egress would be given, either by virtue of the contract documents or by subsequent agreement.

4.5.11.9   Alternatively, in the absence of any such specific warranty, the event could be construed to have been initiated after it was reasonably necessary for the contractor to have such possession, access or egress in relation to their published working schedule illustrating the proper commencement and continuance of the work.

4.5.11.10   **Failure to obtain consents** – As with a failure to provide access, egress and possession by a fixed date, a failure to obtain any third-party consents necessary for the works is initiated on the day after it was warranted that such permissions would be in place, either by virtue of the contract documents or by subsequent agreement.

4.5.11.11   Alternatively, in the absence of any such specific warranty, the event will be initiated after it was reasonably necessary for the contractor to have such consents in relation to their working schedule illustrating the proper commencement and continuance of the work.

4.5.11.12   **Suspension of the works** – If the works are suspended by an instruction to that effect, the instruction takes effect when it is issued.

4.5.11.13   A power to suspend the carrying out of the works following a failure to make prompt payment is a relatively new addition to the risks borne by the employer in relation to delay arising from the provisions of the Housing Grants Construction and Regeneration Act. This Act entitles the contractor to suspend the performance of the work if they are not paid promptly and in full, in accordance with the conditions of contract. For these purposes, the event is initiated on the date the contractor exercises its right to suspend the performance of the works.

4.5.11.14   **Other occurrences** – In relation to other breaches of contract comprising, for example, any delay, impediment or prevention caused by, or attributable to, the employer, the employer's personnel or the employer's other contractors, the date of initiation of the event is the date the breach is committed.

4.5.12   **Acts or omissions of third parties**

4.5.12.1   Identifying the initiation date of acts or omissions of third parties is relatively straightforward as they are of no relevance to the contractor unless they actually affect the contractor's works in some way.

4.5.12.2   To this extent the principles are similar to those on which the timing of a nomination of a subcontractor is construed. Civil commotion, strike or lockouts, for example, are not intervening events at the employer's risk unless and until they affect the progress of the works. In circumstances such as these, the initiation date is that on which the delay to progress actually commences as a result of the event.

4.5.13   **Neutral events**

4.5.13.1   Neutral events are those which, under the contract, are at the employer's risk as to time but at the contractor's risk as to cost. Often these comprise such things as adverse weather, force majeure, labour strikes, and so on. As with the effect of events caused by others, unless they actually affect the progress of the works, they are of no interest. The occurrence of these events can sometimes be predicted a short time before they are initiated but for most purposes, the timing and duration of the effect on the progress of the works cannot properly be predicted and will be a matter of record of what actually happened.

4.5.14 **Disruption**

4.5.14.1   Disruption results in a loss of efficiency and is usually manifest in increased costs for a given amount of work. It may be caused by delay to progress and it may cause delay to progress. It is often complicated by the fact that disruption to a single activity over a short time span can be caused by more than one event at the risk of the contractor and also at the risk of the employer. However, it can be broken down into its constituent parts, retrospectively, by reference to the progress records.

4.5.14.2   Depending upon the circumstances, the evidence of lost efficiency is derived from a comparison between either:

- the resources and productivity planned compared with the resources and productivity actually achieved; or

- the resources and productivity actually achieved during an undisturbed period compared with the resources and productivity achieved during a disturbed period;

- the calculation depends upon what was achieved when the work was not adversely affected by the intervening event being investigated. This can be calculated from the progress records of the history of the activity or activities in question by extrapolating the cumulative resource and productivity data from the progress records or, in simple cases, by a comparison between the planned resources and productivity against the actual resources used and productivity achieved.

4.5.15 **Calculating the effect of intervening events**

4.5.15.1   The effect of an intervening event is calculated by reference to the effect, if any, which the event has on the planned future conduct of the work.

4.5.15.2   Where the event consists of a suspension of the works, in whole or in part, provided that the appropriate calendar is defined, it is satisfactory simply to suspend the working calendar for the appropriate activities for the appropriate period and recalculate the critical path. This is a technique that is particularly useful for dealing with short suspensions (for example, those caused intermittently by bad weather), which, whilst they may affect the whole of the works, may also, in some circumstances, affect only a few activities, which can be reallocated to a specific calendar for that purpose.

4.5.15.3   Where a fragnet is used, the effect of the intervening event is calculated by adding the fragnet to the working schedule, making the appropriate logical connection between the fragnet and the affected activity or activities and recalculating the critical path.

4.5.15.4   The impact of the event is calculated by reference to its effect upon the working schedule at the time of its initiation. This is the process known as 'time-impact analysis'.[10]

4.5.15.5   The feature of time-impact analysis which distinguishes it from progress monitoring is that the former demonstrates the effect of discrete events on progress and on completion, whereas the latter merely identifies slippage against a target, irrespective of the cause or criticality.

4.5.15.6   There are two processes by which the method can be used: one which requires the schedule to be updated and impacted to the initiation date of each event sequentially, irrespective of reporting periods; and the other which requires all

---

[10] See the *Delay and Disruption Protocol* published October 2002 by the Society of Construction Law.

events within a reporting period to be impacted sequentially upon the last updated schedule. The latter is more common on complex projects and is commonly known as the 'windows' process or 'contemporaneous-period analysis',[11] each update period being the 'window' in time which is impacted and reported upon.

4.5.15.7   The process to be followed requires first that the network is updated and rescheduled to identify the calculated effect of progress at the data date. The events initiated after the data date (but before the next data date) are then impacted in chronological order of initiation date, and the critical path recalculated after each event. This indicates which events will have any effect in relation to the effect of any other and calculates that effect upon the timing of future activities and the critical path. It is a prospective method, which predicts the likely effect of the event on the planned sequence at the time of initiation.

4.5.15.8   The process is the same irrespective of however many key dates or sectional completion dates are to be considered and whether or not the effect upon any one or more subcontracts or other contractors is to be calculated. However, for ease of interpretation, and applicable to whatever is to be considered (subcontracts, contractors, key dates or completion), it is important that the dates to be considered are flagged with milestones, which can be coded and independently organised and filtered for reporting purposes (see 'Milestone monitoring' in paragraph 4.6.8).

4.5.15.9   Prior to commencing any impacting, the updated working schedule should be copied to another so that, prior to update, the schedule can be archived and the impacted schedule can then be maintained as the new working schedule.

4.5.15.10   On the addition of each event, the critical path should be recalculated and recorded and the resultant changes to the critical path noted and recorded in the planning method statement.

4.6   Progress monitoring

4.6.1   Monitoring of progress is the process of identifying, in relation to a target, whether progress achieved has met, exceeded, or failed to meet, the target.

4.6.2   Progress monitoring has the useful function of facilitating the detection of trends and facilitating high-level reporting. However, unless change is accommodated by rebaselining, in a complex project in which change is inevitable, it will be difficult, if not impossible, to deduce anything meaningful from a comparison between the baseline and progress achieved.

4.6.3   Progress monitoring without rescheduling the critical path (e.g. earned value management/jagged line) should not be adopted as the sole method of managing time in a complex project because it has two significant shortcomings:

■   it cannot usually distinguish between critical and non-critical activities, or detect shift in criticality; and

■   it is generally impossible to predict the effect of a discrete intervening event, simply because it cannot distinguish between the effects of different events occurring over the same time frame, nor can it distinguish between the slippage caused by events at the contractor's risk and by those at the employer's risk.

4.6.4   There are a number of accepted methods of progress monitoring, the most common methods being:

---

[11] Ibid in Table 1.

- target schedule (rescheduled compared to baseline)

- jagged line

- count the squares

- milestone

- cash flow

- earned value

- resource

4.6.5 Target schedule

4.6.5.1   This method is not possible with some scheduling-software products as it requires the facility for illustrating two or more schedules at the same time on an activity-by-activity basis. The sort of display envisaged is illustrated in Figure 35.

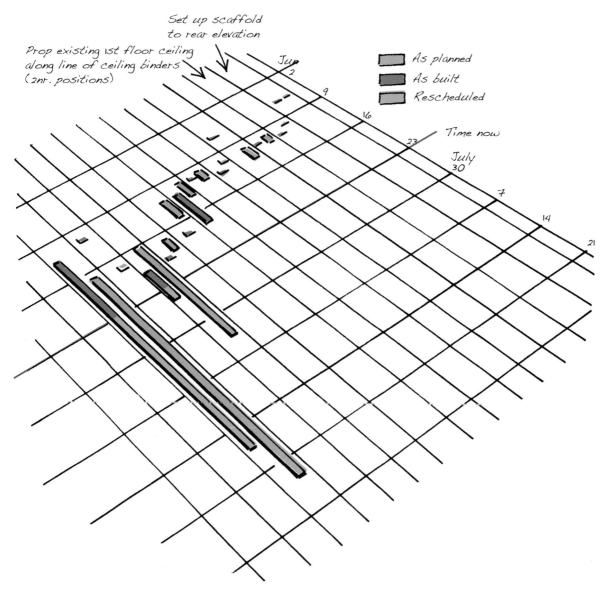

Figure 35 Target schedule comparison.

4.6.5.2   Essentially, the method requires the scheduler to:

■   update the working schedule with progress achieved;

■   reschedule it to a given data date and recalculate the critical path;

■   if no changes have been made, call up, by way of a target, the initial working schedule; if changes have been made, call up the working schedule updated to the previous data date.

4.6.5.3   This is the simplest and potentially most useful method of monitoring progress and compares what was formerly intended as the construction schedule against what actually happened. It is, typically, reported in a bar-chart format along with a narrative to explain why the former planned intent was not achieved and why activities or sequences became delayed and prolonged.

4.6.5.4   It is a useful method for mapping any departure from a previous intent and, if used in conjunction with filtering, can be used to illustrate such things as a shift in timing of subcontract works and change in the critical path.

4.6.5.5   Because both the target and the working schedule are critical-path networks, this method also provides the necessary information for tracing the cause of departure and for identifying what can be done to correct it.

4.6.6   **Jagged line**

4.6.6.1   This method is similar to the target schedule except that it compares the difference between the planned and actual progress on one schedule (by drawing a vertical line that is jagged) as and where progress deviates from that planned (usually commencing at the top of the page on the data date/progress date).

4.6.6.2   However, this method is sometimes used with static schedules in which the activity dates remain constant and progress is marked up against the original intent. In this format, it is useless for time management. In order to function as a useful tool, the effect of progress achieved must always be identified by rescheduling to a data date, with all work done to the left and all work to be done to the right of the data date.

4.6.7   **Count the squares**

4.6.7.1   Progress monitoring by the 'Count the squares' (CTS) method[12] is a technique that has been used for many years as a means of assessing progress for the purpose of high-level reporting in simple projects but it is of little use in complex projects.

4.6.7.2   The CTS method is essentially a bar-chart monitoring tool in which the information provided is only a superficial indication of progress as a whole, rather than specific to a particular activity.

4.6.7.3   In CTS, the work content of the activities is taken to be the same unless subjectively 'weighted' and this tends to produce results which are likely to be unreliable.

4.6.7.4   CTS monitoring will not identify what has caused a departure from required progress, the effect upon the critical path or paths, nor what can reasonably be done to recover from a predicted delay to completion.

4.6.8   **Milestone monitoring**

4.6.8.1   Milestone monitoring is usually used in conjunction with other monitoring processes to identify predicted slippage in significant events or deliverables.

---

[12] See *Project Sponsorship: Planning and Progress Monitoring*, Guidance Note No. 7, The Central Unit on Purchasing, HM Treasury (1986).

4.6.8.2 Whilst it is self-evident that a milestone can never be partly achieved, unless it is closely defined, it may be a matter of opinion whether or not achievement has in fact been accomplished by a given date. In order to achieve a common understanding, milestone definitions should be established, setting out what the state of the works must be in order for the milestone to be deemed completed. It should also be logically connected to the last activity planned to be completed before the milestone can be considered to have been achieved.

4.6.8.3 Provided they are logically linked to the appropriate predecessor and successor activities, when properly reviewed, revised, updated and impacted, the schedule will reflect the dates upon which the milestones are likely to be achieved.

4.6.8.4 A schedule filtered to illustrate only the milestones is known as a milestone schedule.

4.6.8.5 Typically, milestones may be adopted to mark such things as:

- completion of user requirements and sign offs;

- design stages and interfaces;

- applications for licences and permissions;

- key dates for 'topping-out', 'weatherproof', or 'power-on', or other contract requirements;

- a payment stage;

- a sectional completion date;

- the start and finish of a subcontractor's work, or phase of work;

- delivery and removal of key personnel, facilities, equipment, or other important resources;

- handovers and interfaces with key suppliers, statutory authorities and third parties;

- significant dates for receipt or delivery of approvals and information.

4.6.9 Cash-flow monitoring

4.6.9.1 This method of progress monitoring relies entirely upon financial values. Essentially, the method requires a planned cumulative cash-flow forecast and the cumulative certified value and/or costs to date.

4.6.9.2 Monitoring is achieved by a comparison between the two, usually by cumulative graph against valuation dates. The departure between the lines of the graph, if any, will indicate by reference to the valuation dates approximately when the departure occurred (but not the cause of it) and, by extrapolation to the predicted total value at a predicted expenditure rate, it can give an indication of when completion is likely to be achieved and the works' ultimate value.

4.6.9.3 However, cash-flow monitoring will not identify what has caused a departure from required progress, the effect upon the critical path or paths, nor what can reasonably be done to recover from a predicted delay to completion.

4.6.10 Earned-value management

4.6.10.1 Earned-value management (EVM)[13] is a practice based on earned-value analysis (EVA). It is a cost-monitoring rather than a time-monitoring tool. It is a

---

[13] For a detailed explanation of EVA, see BS6079-1:2002; AS4817:2006; ANSI748B.

more sophisticated method of progress monitoring by value than cash-flow monitoring as it attributes actual rather than notional values to the activities either by reference to the financial value or the labour/plant-resource value, or both. Provided that the productivity achieved in the valuation period is as precisely calculated, this can give a useful illustration of the progress achieved against what was planned to be achieved by reference to the earned value of the comparable work.

4.6.10.2   EVM is established by allocating a budget to the plan (usually this is in terms of money and man days). From that data a cumulative-expenditure graph can be drawn for the project overall and for elements within the WBS, known as the planned value.

4.6.10.3   The two most useful key-performance indicators and the basis for calculations of future performance are the schedule-performance index ('How are we doing against plan?') and cost-performance index ('Are we efficient?').

4.6.10.4   The formulae for these two calculations are:

$$Cost\ performance\ index = \frac{Earned\ value}{Actual\ cost\ to\ date}$$

$$Schedule\ performance\ index = \frac{Earned\ value}{Planned\ value}$$

4.6.10.5   Provided that the work content and the sequence of the schedule used as the baseline are the same as the current schedule, then the benefits of EVA will be:

■   identification of what work has been achieved against the plan; and what it has cost to reach that level of achievement;

■   whether the work achieved has been achieved efficiently (i.e. represents good value for money);

■   whether the project is likely to finish on time; and

■   whether the project is likely to finish on budget.

4.6.10.6   As with all other project-monitoring processes, if a different work content or sequence of working from that planned is adopted, a meaningful comparison between planned and actual will be difficult, if not impossible.

4.6.10.7   EVA cannot identify what has caused a departure from required progress, the effect upon the critical path or paths, nor what can reasonably be done to recover from a predicted delay to completion.

4.6.11   Resource monitoring

4.6.11.1   Resource monitoring can be carried out against labour, plant and/or materials. The principles are the same as in other methods of progress monitoring in that there must be a meaningful target of like content and sequence against which current values can be compared.

4.6.11.2   With resources the comparison can be project-wide or more discretely focused upon a particular zone, element or subcontractor if required. The only prerequisite is a target against which actual resources employed can be measured against the resources planned to be employed.

4.6.11.3   Data will usually be contained in daily-allocation data-input sheets, which can be summarised, and graphs can be produced to show actual resources used compared to those planned and, by extrapolation, predict consequences.

However, resource monitoring will not identify what has caused a departure from required progress, the effect upon the critical path or paths, nor what can reasonably be done to recover from a predicted delay to completion.

4.6.12 **Acceleration and recovery**

4.6.12.1   Irrespective of the causes, or whose responsibilities they are, at some point in every complex project, it will be necessary to consider the recovery of a delay to progress, or acceleration.

4.6.12.2   The strategy for dealing with recovery and acceleration should be a matter of policy and set down in the planning method statement (see Part 2 – Strategy). In principle, the options are usually changes to duration, sequence, method or resources and may include one or more of the following planned processes:

■   design

■   standards of workmanship or materials

■   method of working

■   quantity of work

■   sequence (changes in interfaces/logical dependencies)

■   resources

■   working time

■   contingencies

■   information flow

4.6.12.3   In any particular case, whichever method is to be adopted, its appropriate application will depend upon a number of factors including:

■   at whose risk is the delay to progress?

■   what portion of the works has it affected?

■   what trades or activities has it affected?

■   how much time is to be recovered?

■   on what path or paths is that time to be recovered?

■   what are the activities on those paths?

■   can different methods be applied to different activities and paths?

■   what cost is likely to be incurred?

■   how effective is the recovery likely to be?

■   what are the risks in implementing the selected method or methods?

4.6.12.4   The priority to be given to the various choices will largely depend upon when, in relation to the plan, the recovery or acceleration is to be attempted and the degree of time to be made up. Whatever strategy is adopted, the answers to the above issues and hence the reasoning for them and justification should be clearly set down in the planning method statement consistent with the reviewed and revised working schedule to which the strategy is to be applied.

4.6.12.5   It should always be borne in mind that whilst contingencies will be readily identifiable as part of the schedule activities and logic, they should be preserved as long as possible consistent with the current risk appraisal. If contingencies are absorbed in the early stages of a project, the likelihood of completion on time will be reduced. This is simply because, as the project proceeds, the number of other alternative methods available to make up time will naturally reduce (see Part 2 – Strategy).

4.6.12.6   Accordingly, contingencies should be reviewed and revised in the light of the associated status of the risks and the working schedule. When reviewing the contingencies, review and revise time allowances for those activities concerning contractors' and employer's contingencies independently, depending upon the party responsible for the effects of the intervening event which caused delayed progress to be required to be recovered.

4.6.12.7   Risk for which a contingency is provided, and which then matures, will be accommodated in the schedule as either one activity, or a series of activities added to it, which then replace (in whole or in part) that contingency.

4.6.12.8   Where it is evident that the anticipated risk has not matured to the extent provided for, or at all, the contingency will be reduced or omitted entirely and replaced by the as-built series of activities as they actually occurred.

4.6.12.9   Redundant contingency should be either added to later contingencies (as a result of reappraisal of enhanced risk) or omitted entirely.

4.6.12.10   Prior to commencing the planning of any recovery or acceleration, the working schedule should be copied to another so that, prior to revision, the schedule can be archived and the revised schedule can then be maintained as the new working schedule.

4.6.12.11   During revision, an audit trail should be maintained of the changes made, and the reasons for them, recorded in a revision of the planning method statement.

4.6.12.12   On completion of all revisions, the critical paths should be recalculated and recorded and the resultant changes to the critical paths noted and recorded in the planning method statement.

5 Communicating and integrating the time-model

5.1 Introduction

5.1.1   Reporting is an extremely powerful tool which enables key stakeholders to make informed decisions, with regard to time management, in an efficient and effective way. Good reporting ensures that all parties have the same perspective on the status of the project.

5.1.2   A well-structured report enables the users to easily extract key information pertinent to their role.

5.1.3   Reporting is the process of disseminating the required information to those who want it, need it, or should be informed. Necessarily, it involves communication of a message which, as in any other form of communication, must be received and understood if it is to be acted upon.

5.1.4   Whether the report is required to be delivered according to a calendar period (e.g. following the occurrence of an intervening event), or regularly (irrespective of events), or occasionally (depending upon the nature of the data) goes to the root of the significance of the report. What the contract requires in that regard should be closely studied in relation to the purpose of the report and the parties to whom it is to be directed.

5.1.5   This is important because, under some forms of contract, anything issued outside the required time, or not to all the specified parties may have the contractual effect of shifting the liability for the consequences of certain events from the contractor to the employer, or vice versa.

5.1.6   The form in which the report is to be issued may be stipulated under the contract but, in most circumstances, all that is usually required is that it is unambiguously issued on a particular date to a particular party. For those purposes most reports are issued as documents in one form or another and, if issued orally, are confirmed in writing.

5.1.7   However, whilst being issued as a document infers that it will be in writing, for the purposes of time management, because the facts are usually complex, figures, bar charts, graphs, histograms and sub-networks may also be usefully employed in the communication, depending upon the subject matter and target.

5.1.8   At management-board level, simple figures, charts and graphs are extremely useful in communicating complex data. Animations can also be extremely useful in communicating complex procedures and sequences. At site level, however, the devil is in the detail and, accordingly, raw data is often significantly more useful.

Guide to Good Practice in the Management of Time in Complex Projects, by The Chartered Institute of Building
© 2011 The Chartered Institute of Building

5.2   Report types

5.2.1   Reports can be directed to two different types of recipients, external and internal, and can fall into at least three categories.

5.2.2   External recipients include the employer, design team, financiers and such parties as permitting agencies.

5.2.3   Internal recipients will include the contractor, construction manager, project team, site superintendents, contract administrators, subcontractors and suppliers.

5.2.4   The reports generally fall into the following categories:

■   contractual

■   managerial

■   informational

5.2.5   **Contractual notice**

5.2.5.1   Most forms of contract require the contractor to give notice of the occurrence of a notifiable risk or event to the contract administrator (for example, finding ground conditions which are not as described in the contract), which causes delay either to progress of an activity alone, or to a sequence of activities, or to completion, in one form or another, within a limited period following the occurrence.

5.2.5.2   For the precise requirements under the contract, reference must be made to the contract in question. However, in general, where time management is concerned, whereas the 'notice' itself may require little more than the formality of advice that a specified event has occurred and that it is likely either to cause, or has caused, delay either to progress of an activity alone, or to a sequence of activities, or to completion, much more is needed for management of the risk.

5.2.5.3   Because the requirements of the initial notice are often fulfilled by a notice containing very little useful information, most forms of contract also call for later, supplementary information to be provided, concerning the extent of delay to progress and the extent of delay to completion likely to be caused. This is usually to be provided within a limited period after the initial notice has been served. If the contract specifies how this information is to be provided, then that must be complied with.

5.2.5.4   However, in the absence of contractual requirements, the supplementary information should contain all that is necessary to identify:

■   the nature of the event and its unique identifier;

■   the contractual clauses under which the notice is required;

■   a synopsis of the event and the timing of the constituent activities;

■   the planned activity or activities affected and the manner of the effect;

■   the likely consequential effect on the key dates and completion date, if any;

■   the likely consequential effect on any other contractors and subcontractors, if any; and

■   the file reference of the impacted working schedule.

5.2.5.5   The textual information in the above items is supplemented most effectively and usefully by a picture of the fragnet used to impact the event, if relevant, together with a trace of the critical paths from the event to the relevant completion dates on the impacted schedule.

5.2.5.6   In the example of a notice of delay in Appendix 3, the event description is: 'Instructions to resolve a discrepancy between design information'. It is expressed that way because that is the phraseology used in the contract in the example to describe the event which entitles the contractor to an extension of time and compensation. The effect, 'delay to the MOF Jetty', is not the event. That is the direct effect of the event.

5.2.5.7   For clarity, the direct effect should be clearly differentiated from the causative event giving rise to it and, if relevant, the consequential effect on any contractors, subcontractors, key dates, sectional completion dates, or the completion date flowing from the direct effect.

5.2.6   Managerial reports

5.2.6.1   This type of report may be required by the contract, in which case the contract may also set out the specific requirement. Additionally, the project-execution plan should record the reporting requirements, indicating to those concerned the content, form and detail of the various managerial reports required under the contract.

5.2.6.2   The purpose of managerial reports, as the name suggests, is to facilitate the effective management of the project. In this Guide we are particularly concerned with the management of time and for that purpose, the managerial reports are usually referred to as 'progress reports'.

5.2.6.3   Essentially, for the purposes of time management, the progress report will contain information set out under the following headings:

■   executive summary;

■   in the last reporting period;

■   in the medium-term and long-term future; and

■   in the next reporting period.

5.2.7   Executive summary

5.2.7.1   This should provide a high-level summary of the project status and key issues affecting the project, including the effect upon:

■   key dates

■   other contractors

■   subcontractors

■   employer's deliverables

■   sectional completion dates

■   the completion date

■   work trends

■   effect on risk

5.2.8   In the last reporting period

■   what was the high-density working schedule?

■   what actually occurred and when?

■ what caused the difference?

■ what is the likely effect of any difference upon:

    ○   key dates?

    ○   other contractors?

    ○   subcontractors?

    ○   employer's deliverables?

    ○   contingencies?

    ○   sectional completion dates?

    ○   the completion date?

5.2.9 In the medium-term and long-term future

■ what information has been received to affect the medium-density and/or low-density schedule?

■ what recovery or acceleration will be required (if any)?

■ which contractors' contingencies will require adjustment and to what extent?

■ what (if any) risks may occur?

■ what is the strategy for dealing with those risks?

■ which employer's contingencies will require adjustment and to what extent?

■ in summary, what recovery or acceleration (if any) is planned and how is it to be implemented?

5.2.10 In the next reporting period

■ what is the high-density schedule for the next three months?

■ in the high-density schedule for the next reporting period:

    ○   which contractors and subcontractors are affected?

    ○   what planned resources and productivity are assumed?

    ○   what employer's deliverables will be required and when?

■ what (if any) risks may occur, or become greater, or lesser?

■ what is the strategy for dealing with those risks?

■ which contingencies will require adjustment and to what extent?

■ what recovery or acceleration (if any) is planned and how is it to be implemented?

5.2.11   Progress reports will usually be required to be accompanied by reports upon contractual key dates or sectional completion dates; each should be treated as a separate critical path and reported upon independently of the other.

5.2.12 Informational reports

5.2.12.1   This type of report may be required by various stakeholders, in which case the specific requirements will be set down in the project-communication plan for the report. However, in general, where time management is concerned, informational reports will usually include financial control, quality control and testing, health and safety provisions and accidents, data flow in drawing release, requests for information, change control and so on. Much will depend upon the type of data to be transferred as to what is required to transfer it effectively.

5.3   Reporting formats

5.3.1   The form of presentation depends upon for whom the report is prepared and a different process and output will be required depending upon whether the report is needed for strategic or operational purposes and whether the recipient is to make a decision, give advice, monitor progress or simply receive information. However, for most purposes, it will be helpful to have an executive summary at the beginning of any report, highlighting the important points with a cross-reference to the detail. In that way, even the busiest of people can quickly grasp the principles of what is being said and easily drill down to the detail, where they need to.

5.3.2   A good general principle to keep in mind when thinking about the best way to present information is that at site level, reports will usually be required with precision and detail whereas at executive level, summaries with trends and exceptions tend to be more easily digested.

5.3.3   Whenever tables of data are necessary, visuals should always be considered as a method of summarising the effect of that data. Typical visual methods are:

- histograms

- pie charts

- graphs

- PDM network

- ADM network

- bar charts

- linked-bar-chart network

5.4   Feedback and benchmarking

5.4.1   If the principles of this Guide are followed, every project will provide a plethora of data from which many uses can be made in the improvement of performance in future projects.

5.4.2   The data will have been captured throughout the life of a project by reference to the progress-records database, the updated working schedule and the impacted schedules, and these can be used effectively to identify any departure between the investigated state and the desired standard in any particular case.

5.4.3   There are two aspects to benchmarking. On the one hand, the process will establish achieved norms of productivity and activity durations for common and project-specific working conditions, work types, trades, resources and so on and, on the other hand, for common data it will provide some degree of comparison between the time-model performance and industry best practices or standards.

5.4.4 **Benchmarking procedure**

5.4.4.1   The first step in any analysis of performance data is to identify what is to be analysed. This should be done before construction begins so that the planned work type can be identified for tracking, use, or production or retrospective analysis. The more closely that it can be defined, the more easily any analysis required can be performed and the more useful will be the data.

5.4.4.2   For example, in order to identify the normal productivity rate for given work types, the process will normally adopt the following procedure:

■   Identify the various activities comprised in the work and examine the relationship between them. If there was a reasonable degree of continuity between relevant activities, then the data can be taken together to represent the performance of that work type from beginning to end. If not, then the analysis will also provide information on the degree of productivity lost as a result of intermittent working (if any) by comparison between the best continuity achieved and that achieved during broken periods, in the absence of any other affecting events.

■   Identify the resources used throughout the various activities and establish whether there is any significant fluctuation between them in continuous and discontinuous operating conditions. If the resources were constant, then whatever fluctuation in productivity is apparent will have been caused by something other than resources. The effect of any known event, such as the effect of a learning curve, or an intervening event, can then be isolated and filtered out and the remainder will represent the average normal achieved productivity for that work type. Where there are remaining fluctuations, the best achievable productivity and average productivity can be established relatively easily.

■   Where there are fluctuating resources, then a separate analysis of achievable productivity should be made against each combination of resources to establish the effect of the different combinations on the average and best productivity in normal conditions.

■   Where commonly occurring events have interfered with productivity, data for the effect of those on the selected work type can also be established on differing combinations of resources.

■   Tabulate the results and the criteria adopted for identification of each separate analysis. Depending upon the activity type, the productivity will be measured in a quantity of work against a fixed time period, for example, cubic metres per day or linear metres per hour.

5.4.4.3   The effect of different conditions on performance, and the circumstances under which the work was carried out, are essential to the usefulness of the analysis of past performance on future planning, and for the best use of such records, the data should be kept in a database, which can be searched, organised and filtered for future use. Typically, in relation to any search, the data retrievable should consist of values in at least the following data fields:

■   job name;

■   job type;

■   country and region;

■   date started, completed and construction period;

- design team;

- project-management team;

- construction-management team;

- activity type and whether it is a common activity, or with project-specific difficulties;

- activity duration of which data is measured;

- best working conditions' characteristics;

- best working weather conditions;

- best productivity achieved;

- average uninterrupted productivity achieved;

- average working conditions' characteristics;

- most productive resource combination;

- average resource combination;

- effect on productivity of specific events (by type, e.g. multiple reissues of drawings).

5.4.4.4   For some data, such as working conditions, a pick list of possibilities might be established to restrict data input to available values for searching. Once the data has been accumulated, then comparison can be made with other known data to establish whether there is anything inherently out of the ordinary. For example, comparison with common activity types can be made against data in trade manuals.

5.4.4.5   Benchmarking requires a high degree of analytical capability. It can be a time-consuming and costly exercise to complete by those unfamiliar with the research necessary to arrive at meaningful data and, unless carried out to a high standard against a quality-assured process, the results can be rendered useless. Accordingly, whenever in-house departments are not maintained for this purpose, consideration should always be given as to whether the benchmarking could be most successfully and effectively tabulated by external consultants, experienced in the process.

APPENDICES

APPENDIX 1   Time risks which may be borne by the employer

1.1   The generic intervening events, the time effect of which is at the employer's risk under some standard forms of contract, are listed below.

1.2   Under some contract forms, a risk that is ordinarily the employer's passes to the contractor if the occurrence could reasonably have been foreseen by an experienced contractor at the time of tender, or if the event is consequent upon any omission or default by the contractor.

1.3   The risks peculiar to a project will be set down in the contract conditions in one form or another and, in determining the nature of the risk, it is important that attention be paid to the precise phraseology used.

1.4   It is important to be familiar with these risks in order properly to appraise the need for contingencies in any particular case, for both the employer and contractor.

1.5   Subject to the terms of the particular contract, the contractor may be entitled to:

- compensation for a delay to progress; or

- an extension of time to complete; or

- an extension of time to complete and to compensation for prolongation arising out of the following intervening events:

1.5.1   Instructions given to:

- correct any ambiguities or discrepancies in the contract documents;

- investigate and report upon the practicality of alternative value engineering proposals;

- overcome unforeseen physical conditions or artificial obstructions;

- correct any impossibility or illegality;

- rectify any error in setting out caused by incorrect data supplied by the design team;

- carry out exploratory work;

- ensure conformity with any Act of Parliament, regulation or bye-law;

- conform to any condition or limitation in any licence obtained after the award of the contract;

- dispose of human remains, fossils, coins, articles of value, or antiquity and structures, or other remains, or things of geological or archaeological interest;

- make any tests not described in the contract in sufficient detail for the contractor to have allowed for it in his tender;

Guide to Good Practice in the Management of Time in Complex Projects, by The Chartered Institute of Building
© 2011 The Chartered Institute of Building

■ suspend the progress of the works or any part of the works;

■ order any variation;

■ order a change in the employer's requirements;

■ expend a provisional, prime cost or contingency sum;

■ make good the failure of a nomination;

■ amend or alter the health and safety plan.

1.5.2   Variations arising from the supply to the contractor of modified drawings, or further drawings or instructions.

1.5.3   Any failure or inability of the contract administrator to issue, at a time reasonable in all the circumstances, drawings, specifications or instructions necessary for the carrying out of the contract works.

1.5.4   The discovery of adverse physical conditions or artificial obstructions.

1.5.5   Any failure to consent in due time to a contactor's submittal.

1.5.6   The application of unreasonable conditions to consent to a contractor's submittal.

1.5.7   The rectification of loss or damage arising from any of the insured risks.

1.5.8   A failure to obtain any required consent for the permanent works.

1.5.9   Interference caused by any contractors employed by the employer.

1.5.10   Interference caused by any other properly authorised authority, or utility, or statutory undertaker in connection with their statutory obligations, or failure to do work.

1.5.11   The act, or omission of a nominated subcontractor and/or nominated supplier.

1.5.12   The discovery on the site of human remains, fossils, coins, articles of value, or antiquity and structures, or other remains, or things of geological or archaeological interest.

1.5.13   The uncovering, making openings in, or through, reinstating and making good any part or parts of the works covered, which are found to have been carried out in accordance with the contract.

1.5.14   A failure to give access, egress, or possession.

1.5.15   A failure to provide anything which the employer is to provide.

1.5.16   Any increase or decrease in the quantity of any work as a result of the quantities exceeding or being less than those stated in the bill of quantities.

1.5.17   The correction of any error in description in the bill of quantities.

1.5.18   The wrongful withholding of consent to the expulsion of a nominated subcontractor, pursuant to any forfeiture clause or rescission of the subcontract.

1.5.19   Any suspension by the contractor of the performance of his obligations for a failure to make payment.

1.5.20   A variation to the health and safety plan.

1.5.21   Any delay, impediment, prevention or default by the employer.

1.5.22   Exceptionally adverse weather conditions.

1.5.23   Other special circumstances of any kind whatsoever that may occur.

1.5.24   Force majeure.

1.5.25   Strike, lockout, or combination of workmen, civil commotion, strikes or lockout.

1.5.26   The effect on labour, materials or goods of government acts or controls.

1.5.27   Inability to obtain labour, materials or goods, which could not have been foreseen at the date of tender.

1.5.28   An event beyond the contractor's control.

Appendices

APPENDIX 2   Desirable attributes of scheduling software

2.1   Primary considerations

2.1.1   Projects and sub-projects

2.1.1.1   Software which can only cope with a single project at a time is unlikely to be sufficiently flexible for complex projects. For example, apart from the possibility of identifying as sub-projects separate sections (which are subject to sectional completion, or separate key dates), for ease of application in practice, it may also be useful to identify separate operational zones as sub-projects.

2.1.2   Activities

2.1.2.1   For each activity there should be:

- a unique activity-identifying alpha-numeric code;
- a unique description.

2.1.2.2   Software which permits duplication of activity IDs or activity descriptions, without warning, is likely to produce schedules which lack clarity and are thus incompatible with good practice. The software should not facilitate that duplication at all, or, if it does, have a clear, permanent warning on the schedule as to the deficiency.

2.1.2.3   The software should be capable of distinguishing between the following activity and event types:

- duration-identified activities;
- resource-calculated activities;
- hammocks;
- start milestones or flags;
- finish milestones or flags;
- employer-owned contingency/risk allocation;
- contractor-owned contingency/risk allocation.

2.1.2.4   An activity-related field capable of taking free text and numbers as comments or notes is often a useful facility.

2.1.2.5   The software should be capable of identifying activity durations in different formats. Although for most purposes in construction, activity durations in days may be sufficient, for the purposes of limited possessions, durations in hours and minutes and in outline schedules, durations in weeks and months are necessary.

2.1.2.6   The software should make clear to what unit of time it carries out its calculations (i.e. days, hours, minutes or seconds). The best software calculates to the minute.

2.1.2.7   The software ought to be capable of identifying which activities are logically determined to be of a shorter duration than the applied logic and whether they are to be 'stretched' or 'not continuous' as a result of the logic.

Guide to Good Practice in the Management of Time in Complex Projects, by The Chartered Institute of Building
© 2011 The Chartered Institute of Building

Appendices

2.1.3 **Logical relationships**

2.1.3.1   The software should permit a logical flow of work and prohibit the indication of relationships which are impossible to perform in practice. It should provide forward and backward passes and detect loops and open ends. Any software which fails to do this is likely to produce schedules that are incompatible with good practice. It should not facilitate such defects at all, or declare a clear, permanent warning on the schedule as to the deficiency.

2.1.3.2   The software should be capable of identifying all variations of logical links, either individually or in combination. Software which limits the user to finish-to-start logic, or to few logical connections to any one activity, is unlikely to be useful. The software should permit at least two links from the start and finish of every activity.

2.1.3.3   The software should identify any inconsistency between logic and the activity durations to which the logic is applied.

2.1.3.4   Logic should be capable of being illustrated as 'driving' or 'non-driving' to any chosen point within the model.

2.1.3.5   Logic should distinguish between:

- engineering logic (the construction sequence with no resource constraints);

- resource logic (the construction sequence carried out with the available resources);

- preferential logic (the construction sequence with imposed constraints to modify the purely 'engineered' and/or 'resourced' construction sequence);

- logic linking zones and/or sub-projects;

- additionally, the software should be capable of identifying fixed lead and lag and the working calendar the lead or lag is to adopt;

- lead and lag should be listed as logic attributes.

2.1.4 **Constraints**

2.1.4.1   Manually applied constraints are likely to be useful on most projects. Those that are acceptable, when correctly applied, are:

- start-no-earlier-than a given date;

- start-no-later-than a given date;

- zero-free-float;

- additionally, the software should be capable of clearly identifying when a manual constraint has been applied to an activity.

2.1.4.2   Some software facilitates the use of constraints which will manipulate criticality and inhibit the ability of the software to accurately model time. These are not acceptable in a schedule used to manage time. Those which, if available in a software product, should not be permitted to be used without a clear, permanent warning on the schedule as to their effect are:

- any combination of constraints which will fix the earliest and latest dates for any activity or milestone;

- a mandatory start date;

■ a mandatory finish date;

■ zero-total-float.

2.1.5 Critical path

2.1.5.1 The software should be capable of identifying:

■ the longest path to completion;

■ the longest path to intermediate key dates or sectional completion dates;

■ logic and activities which are critical, separately from those which are not critical to one or more completion dates;

■ total float on each path;

■ free float on each activity, on each path.

2.1.5.2 The software should be capable of facilitating the tracing of a critical path or paths through the driving logic of each activity on the critical path to a particular completion date or key date from time to time.

2.1.6 Calendars

2.1.6.1 The software should be capable of facilitating the use of a number of different working calendars for activities, resources and lags, each capable of identifying different:

■ working-week start day

■ working weeks and weekends

■ working days

■ working hours

■ holidays

■ standard calendars and exceptions

2.1.7 Resources

2.1.7.1 The software should be capable of facilitating the use of a number of different resources and determining a realistic schedule based on the resources allocated to each activity.

2.1.7.2 The resource-scheduling facilities may be implemented in numerous different ways and, if resource analyses are to be correctly interpreted, it is essential for the scheduler to be able to understand how any specific tool works. Accordingly, the software documentation should provide comprehensive information on the various algorithms and options available within the tool.

2.1.7.3 The software should be capable of identifying resource availability at project level by reference to:

■ resource ID or unique identifying code

■ name of resource

■ resource working calendar

■ resource availability levels by date or time period

■ normal and maximum overload levels for each resource

■ the cost rates for the resource to be defined in terms of:

○ normally available

○ unused

○ overloaded

○ overtime working

○ lump sum

2.1.7.4   The software should be capable of allocating resources to an activity to permit:

■ the allocation of several resources to each activity;

■ the start and end time for each resource to be nominated in relation to the activity duration;

■ additionally, the software should be capable of identifying a split activity duration caused by resource shortages;

■ the quantity of each resource allocated to the activity to be varied at defined times relative to the activity duration;

■ resources to be allocated as a quantity and distributed over the activity duration based on a nominated profile;

■ the option of either maintaining the activity duration or calculating the duration based on effort (within defined maximum and minimum resource levels).

2.1.7.5   The software should be capable of calculating resources:

■ by progressing from project start and with the facility for:

○ resource smoothing without end-date extension (i.e. by using only the available float to minimise resource overloads when all of the available float has been consumed);

○ resource smoothing with end-date extension (i.e. by using the available float to minimise resource overloads until all of the available float has been consumed and then, when the maximum overload threshold is reached, delay critical activates to keep resource demand at, or below, the maximum overload threshold level);

○ resource levelling without overloading (i.e. by levelling the resource demand to remain at, or below, the planned level by delaying activities and end date if necessary);

○ cost calculation (i.e. based on the resource cost parameters, calculate the cost of each activity and the aggregate cost for the project by time period and in total);

■ by progressing backwards from project finish, using the results of the project-start allocations, with the facility for restraining the pacing resources to the normal or maximum overload level.

2.1.7.6   The software should be capable of resource reporting to demonstrate:

■ resource aggregation (e.g. by summarising each resource by time period based on the activities' early start and late finish dates);

■ resource histograms illustrating the effect of unlimited and planned resource quantities by time period; these should be capable of showing:

    ○   normally available

    ○   maximum threshold

    ○   used

    ○   unused

    ○   overloaded

■ for each activity:

    ○   the controlling resource when an activity has been delayed during the resource-levelling procedure;

    ○   the calculated cost of the resources used on the activity;

■ project-cost reports identifying:

    ○   the cost per time unit of the resources allocated;

    ○   the cumulative project cost (e.g. project costs may include unused resource costs and various contingencies);

    ○   resource reports identifying the activities allocated to each specific resource;

    ○   aggregate reports identifying resource utilisation and cost reports aggregated by a time unit (e.g. month) or by a schedule component (e.g. a WBS element).

2.1.7.7 The software should be capable of permitting resource allocation to be updated to:

■ record actual resource usage and actual costs;

■ adjust planned resource usage on activities in progress and future work without impairing the use of archived previous editions of the schedule.

2.1.7.8 The software should be capable of identifying resource capabilities by reference to:

■ skill type;

■ alternative resources (i.e. if the nominated resource is overloaded, automatically use a nominated alternative);

■ resource-breakdown structures (with aggregate reporting similar to the WBS);

■ stacked histograms;

■ alternative resource-levelling and smoothing algorithms with clear descriptions of their functionality.

2.1.7.9 The software should also be capable of being able to:

■ store resource baseline archives separate from the current schedule;

■ pool resources across multiple projects;

■ manually set activity and project priorities (to influence resource allocations during resource scheduling);

■ pool resources within a project (see alternative resources);

■ perform earned-value aggregations.

Appendices

2.1.8 **WBS and activity coding**

2.1.8.1   The software should be capable of identifying a work-breakdown structure. Whilst a structure of eight levels should be the ideal, a structure of fewer than five levels is unlikely to be practical on a complex project.

2.1.8.2   The facility for a broad variety of bespoke database fields which can be displayed is usually an essential requirement of complex schedules.

2.1.9 **Organisation**

2.1.9.1   The software should be capable of organising the layout in any combination of fields and attributes, sorting activity, logic, attributes and values in any field.

2.1.10 **Filtering**

2.1.10.1   The software should be capable of filtering the content of any layout by selection of the value of any field, or attribute (either alone, or in combination with other fields, or attributes) on the basis of:

- ■ equal to

- ■ containing

- ■ not equal to

- ■ not containing

and, where the fields contain values, in relation to those values there should also be the facility for selection:

- ■ between

- ■ not between

Filtering should also facilitate the use of Boolean 'AND' and 'OR' logical combinations.

2.1.11 **Layout**

2.1.11.1   The minimum available layouts should comprise:

- ■ bar chart without logic

- ■ bar chart with logic

- ■ network diagram (ADM or PDM)

- ■ resource profile

- ■ cost profile

2.1.11.2   The software should have the facility for creating and saving a variety of different combinations of fields and attributes, organised and filtered, as layouts for reporting purposes.

2.1.11.3   The timescale to which the layout is restricted to view should be identifiable to any duration and density during the period between 6 months prior to inception of the earliest project and 12 years after planned completion of the latest project.

2.1.11.4   Every layout should be printable as both hard copy and PDF.

2.1.12 **As-built data**

2.1.12.1   The software should be capable of identifying the factual data for each activity and resource as:

- actual duration

- start date

- finish date

- percentage complete

- remaining duration

- calculated cost

- actual cost

- certified value

- resources expended

- productivity quotient achieved to date

2.1.13 **Updating**

2.1.13.1   The software must be capable of identifying a data date by a straight line through the activity bars at that date.

2.1.13.2   The software must be capable of comparing progress against the currently agreed baseline, such that any delays and/or changes in activity sequencing are clearly demonstrated.

2.1.13.3   The software must be capable of recalculating the critical path or paths and the predicted early and late start and finish dates of all activities and resources against the data date with the effect that:

- all activities indicated to have started or finished are indicated to have started, or finished, earlier than the data date;

- no activity is identified to have started, or finished, later than the data date;

- activities which are in progress at the data date are indicated to be due to finish on a date after the data date proportionate to their degree of progress in relation to their planned duration at the data date.

2.1.14 **Inputting and editing data**

2.1.14.1   The software should be capable of holding input data and edits in memory so that they are subject to 'undo' and only saved on a positive instruction to do so.

2.1.15 **Archiving**

2.1.15.1   Files should be capable of being saved in compressed-data format for archival purposes.

2.1.16 **Training and support**

2.1.16.1   The availability of effective, product-related training is extremely useful even for experienced schedulers. Even with the simplest of software, it is always helpful to understand how the software supplier identifies that it should be used.

2.1.16.2   Because of the sophistication of modern software and the inability or unwillingness of the manufacturers to subject products to rigorous testing before release, committed and easily available software support and continuously updated software is more important today than it has ever been.

2.2   Secondary considerations

2.2.1   Those matters which do not add to the quality of the calculated output but to the manner of use and which, depending upon circumstances, may be of some importance are:

2.2.2   Enterprise-wide software

2.2.2.1   Enterprise-wide software which can directly link a project or projects across the internet such that large, complex projects are able to be scheduled and effectively monitored across the world.

2.2.2.2   Enterprise-wide software which can relate together all projects with which the company is concerned. It is a useful attribute for enhancing company management.

2.2.3   Communications

2.2.3.1   Whether the schedule can be accessed by other parties via the internet, in whole or in part, for viewing only, or for editing with security-dependent access rights, can be of importance in managing the schedule.

2.2.4   Appearance

2.2.4.1   Software capable of being customised according to company requirements for house style by using different fonts, line thickness, or type and colours for each available field, value in the database and/or the background is useful.

2.2.4.2   A drawing facility which can be used to highlight aspects of a report is often useful.

2.2.5   Comparison of schedules

2.2.5.1   For the purpose of identifying the effect of differences between schedules in the process of review, revision, updating and impacting causative events, it is useful to have the facility for comparing two or more schedules on a line-by-line basis. In practice this usually means the facility for identifying one or more target schedules which can be viewed simultaneously with the current schedule.

2.2.6   Organisation

2.2.6.1   A facility for organising the layout in order of logical predecessors and successors is useful.

2.2.7   Transparency with other software

2.2.7.1   The facility for importing from, and exporting to, other scheduling software may be available but if it is, it should be capable of listing the differences which result from such import or export.

2.2.7.2   The software should be capable of importing and exporting from and to other databases such as MS Excel and MS Access for updating, analytical and quality-assurance purposes.

2.2.7.3   The facility for attaching hyperlinks to activity IDs should be available to enable the linking of such documents as photographs and videos, flow charts, procedures and method statements, and progress records.

2.2.7.4   Integration with time-keeping and cost-keeping systems can facilitate automatic updating from time, plant and material records, which, in relation to a fully resourced schedule, can produce an automated update facility.

2.2.8   Risk analysis

2.2.8.1   The facility for stepping through a potential shift in timing of activities to ascertain the consequent shift in the critical path is useful.

2.2.8.2   Monte Carlo analysis will give a profile of likelihood of success against given criteria, which, if accurately predicted against data that remain unchanged, will predict likely outcome.

2.2.9   Archiving

2.2.9.1   A back-up capable of being set to default periods or to be executed manually is a useful facility.

APPENDIX 3   Sample notice of delay

Event Description	Instructions to resolve a discrepancy between design information
Event No.	25
Summary of issues	The walkway design contained a dimensional error in that the dimensions provided did not correlated to the coordinates on the drawing. The walkway steel was ordered from the dimensions provided. The walkway was fabricated and delivered. When it was installed it was short by 0.5 m. The SC amended the design at C's instructions of 20 May, to insert a small panel in mid length to accommodate the error. The walkway structure had first been installed between 1st April and 14th May 1998, however, because the corrections to the dimensional error inhibited its use, completion of the MOF Jetty was delayed until 11th June 1998.
Documents referred to	Letter SC > C 11th May 1998
	Letter SC > C 12th May 1998
	Letter C > SC 20th May 1998
	FI. 51 – 6th June 1998
	Working schedule at 11 May 1998

Sequence

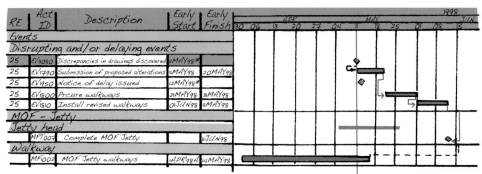

Contact clauses	Clauses 25 and 26
Effect on time	Delay to Sectional Completion of MOF Jetty until 11th June 1998
Effect on cost	Prolongation of costs in late demobilization of sea-borne equipment and prolongation of MOF Jetty

APPENDIX 4 Industry productivity guides common in the UK

Reference may be made to such data as Planning Planet's[14] output-production rates, the Planning Engineers' Organisation's[15] lead times, or other industry standards.

Builders' pricing books will give an indication of durations as well as price for a multitude of different activities, for example, alphabetically:

■ Laxton's[16]

■ POMI[17]

■ Spencer Geddes[18]

■ Spon's[19]

■ Wessex[20]

POMI Building Price Book, rather helpfully, advises the user as to what the authors have relied on in arriving at productivity quotients, saying:

labour constraints are based on feedback from a number of professional estimators, surveyors, tradesmen and specialist contractors. In some instances we have had to take a view where new products have as yet not produced estimating feedback. In the majority of cases site labour times have been carried from real life outputs, reflecting the performance of supervised men working under typical site conditions.

Notwithstanding this reassurance, these standards should be viewed with caution and the user should be careful to compare situations which are inherently job-specific with standard work descriptions and make appropriate adjustments.

Appendices

---

[14] http://www.planningplanet.com/output_page.asp?choice=outputs
[15] http://www.planningengineers.org/knowledge/leadtimes.aspx
[16] *Laxton's Building Price Book,* Butterworth–Heinemann.
[17] *POMI Building Price Book (Principles of Measurement (international) for works of construction),* Barton Publishers.
[18] *Spencer Geddes' Estimating for Building and Civil Engineering Works,* Butterworth–Heinemann.
[19] *Spon's Architects and Builders Price Book,* Chapman and Hall.
[20] *Wessex Engineering Services Price Book,* Wessex Electronic Publishing Ltd.

Glossary of terms

absenteeism	A failure of a labour resource to arrive at the designated work place
acceleration	The making up of lost time at the employer's (q.v.) expense; see also 'recovery'
activity	An identified portion of work
activity-content coding	See paragraphs 3.7.20 and 3.8.21
activity-cost coding	See paragraphs 3.7.21 and 3.8.22
activity description	See paragraphs 3.8.3 and 4.3.8
activity identifier	See paragraphs 3.7.9 and 3.8.2
activity-identifier code	See 'activity identifier'
algorithm	A method of calculation
animation	A computerised reconstruction of a sequence
approximate quantities	The estimated quantity of work, usually prepared from a scheme design and before detailed design is carried out, for the purpose of cost estimating
area	A portion of the work defined for the purpose of management and control
arrow-diagram method	The method of network scheduling in which the task is represented by an arrow and events by nodes; see paragraph 3.4.8
as-built	Work carried out
as-built schedule	See paragraph 3.3.7
as-late-as-possible	See paragraphs 3.8.45.3 and 3.8.46
as-soon-as-possible	See paragraph 3.8.46
audit trail	A sequence of checking whereby data in one document is validated before being relied upon in another
backfill	The material used or the process of using that material to fill an excavation
bar chart	See paragraph 3.4.5
benchmarking	See paragraphs 3.8.13 and 5.4
bid	See 'tender'
bid schedule	See 'tender schedule'
bills of quantities	A document, usually prepared according to defined rules, which sets out the measured quantity of work and describes the quality standard of materials and workmanship for pricing
buffer	See 'contingency'

calendar	See paragraphs 3.7.15 and 3.8.17
cash flow	The balance of money received against money spent according to a defined formula, sometimes referred to as accounting rules; the rate at which money is spent in the past (actual) and in the future (forecast and planned)
CDM regulations	The health and safety rules and regulations applicable in the UK
cell	A unit in a spreadsheet (q.v.) identified by a letter identifying its horizontal position together with a number identifying its vertical position
change management	The art and science of controlling the effect of a departure from the contract quality, quantity, methodology, cost and timing of the work
civil commotion	Usually linked with 'riot' and referring to an uprising of a large number of people
coding structure	See paragraph 3.7.9.2 and Figure 15
completion date	The end-date for the works, the subject of the planning and scheduling process
complex projects	See paragraphs 1.5.1 and 1.5.3
construction manager	The title given to the party responsible for managing trade or subcontractor performance
consultants	Those employed by the employer (q.v.) in an advisory capacity; see also 'design team'
contemporaneous-period analysis	See 'windows'
contiguous duration	The gross period of time required to carry out an activity in an uninterrupted process
contingency	A provision of resource and/or time which may or may not be required
contract administrator	The title given to the party paid by the employer (q.v.) for carrying out duties under the construction contract
cost coding	See paragraphs 3.7.21 and 3.8.22
count the squares	See paragraph 4.6.7
crew	A defined group of labour (q.v.) usually tasked with a single trade (q.v.)
crew size	The quantity of labour in a defined group; see paragraph 3.5.1.4
critical path	See paragraph 3.8.58
critical-path method	See paragraph 3.7.5.1
curing	The chemical process of hardening of a fluid mix of materials
cut and fill	A method of levelling land by 'cutting' from the high parts and using that product for 'filling' the low parts
dangle	See 'open end'
data date	The date upon which a schedule is revised or updated with progress data up to that date; also known as time now
data schedule	An ordered list of activities, their dependencies, durations, resources used and the productivity achieved; see paragraph 3.7.5.2

database	Computerised data storage in which data is recorded as values against fields, which can then be searched and retrieved in a multitude of forms by filter and organisation
database record	The data entered on a database
delivery lead time	The period between placing an order and receiving the goods
density design	See paragraph 3.7.11
dependency	The relationship between two or more activities which determines their respective timing and sequence of operation
design and build	A form of project procurement in which the contractor also carries the design responsibilities
design consultant	A member of the 'design team' (q.v.)
design team	The title given to all those designing any part of the work but not including contractors who carry out design
development schedule	See paragraph 3.3.3
disruption	See paragraph 4.5.14
document-management system	A 'database' (q.v.) with automated input of document-specific values, often with a graphical interface which facilitates the viewing of a scan of the document
domestic subcontractor	A subcontractor chosen by the contractor
double entry	A method of book-keeping in which the data is recorded in different places by duplicating the data input
down time	The period (usually brief) when work is suspended
driving relationship	A relationship by which either the logical start or logical finish of an activity is dependent upon the start or finish of another activity
dummy	A nominal task which requires no work to be carried out in an ADM network; see also paragraph 3.4.8.3; can also be used in PDM to simplify logical connections between activities
duration	The period between the start and finish of an activity
dynamic schedule	A schedule which will react to change and predict the consequences of that change
earned-value management	See paragraph 4.6.10
earthworks	The term applied to describe the 'cut and fill' (q.v.) or regrading of land
effort expended	The amount of work done in relation to the whole
egress	Exit or way out
employer	The party who commissions the work and is responsible for payment under a construction contract; also referred to as 'owner' or 'purchaser'
employer's contractors	Those contractors engaged by the employer (q.v.) (usually for specialised work) under a separate contract from the main contract for the works
employer's goods and materials	Those goods and materials to be provided to the contractor by the employer (q.v.)

end-user requirements	The needs of those who are to use the finished product
engineer, procure and construct	Sometimes called 'turnkey'; a form of contract in which the contractor adopts the obligation of fitness for purpose in designing and providing the finished product ready for use
engineering logic	See paragraph 3.8.27
environmental conditions	Usually weather conditions, but can be other conditions in which activities have to be carried out, e.g. mines and some large-scale, heavy-engineering projects (dams, nuclear power stations and the like)
event register	A listing of intervening events and the salient information concerning those events; see paragraph 3.8.65.5
executive summary	A short version of a longer report, identifying its salient features
expected finish	See paragraph 3.8.48.1
expired time	The period elapsed in relation to the planned duration
filter	An electronic facility for identifying the values in one or more fields amongst others in a database
finish-no-earlier-than	See paragraphs 3.8.47.1 and 3.8.49.1
finish-no-later-than	See paragraphs 3.8.47.1 and 3.8.49.1
finish-to-finish	See paragraph 3.8.33
finish-to-start	See paragraph 3.8.34
flag	A symbol intended to attract attention
flexible constraints	See paragraph 3.8.46
float	See paragraph 3.8.50
float values	The quantity of float on a number of paths
floor slab	A dense construction forming a floor, usually of concrete
force majeure	An intervening event which is outside the control of both the contractor and the employer (q.v.) and for which, unless the contract provides otherwise, the contract is deemed frustrated
formulaic calculations	Calculations produced by formula
formwork	The temporary works required to constrain the pouring of concrete
fragnet	A small network of activities; see paragraphs 4.5.15.3, 5.2.5.5 and Appendix 3
free float	See paragraph 3.8.51
gang	See 'crew'
generic resource	Resource defined by trade or undertaking, e.g. bricklaying
ground beam	A beam linking foundation pads together
guaranteed maximum price	A form of target-cost procurement in which more risk is borne by the contractor than in other forms of contract
hammock	An activity-bar which hangs between the start of the earliest and the finish of the latest in a group of activities, often used for summarisation purposes

health-and-safety planning manager	The party responsible for managing health-and-safety matters ensuring that those responsible comply with health-and-safety legislation
Housing Grants Construction and Regeneration Act	UK legislation providing, amongst other things, for the right to suspend performance in the event that payment is not properly made
impacting	The process of calculating the effect of an intervening event (q.v.) on a dynamic schedule (q.v.); see paragraph 4.5.15
implied variation	An act or omission which is deemed to be a variation
industry standards	See Appendix 4
inflexible constraints	See paragraph 3.8.48
information flow	The transfer of information from one party to another
information-release dates	Dates upon which the design team (q.v.) are bound to release information to the contractor, under the contract
interruptible activity	An activity which, in order to conform to the logic of the network, requires a duration longer than its contiguous duration (q.v.) or which, depending upon the intermittent availability of resources, can be stretched or made longer than the original estimate
intervening event	An event which interferes with the progress of the work; see Appendix 1 for a schedule of those usually at the risk of the employer (q.v.); see also paragraphs 4.5.7 and 4.5.15
jagged line	See paragraph 4.6.6
key date	A term usually used for the date upon which a work stage is to be completed
labor	See 'labour'
labour	The human resource
ladders	See paragraphs 3.8.40.3 and 3.8.71
lag	See paragraph 3.8.36
lagged finish-to-finish	See paragraph 3.8.37
lagged finish-to-start	See paragraph 3.8.38
lagged start-to-start	See paragraph 3.8.39
lead	See paragraph 3.8.36.2
learning curves	The line described by the cumulative graph of productivity in a repetitive cycle as the workforce become more proficient
levelling	See 'resource levelling'
limited possession	A short period in which the contractor is entitled to sole access to a particular part of the work
line-of-balance diagram	See paragraph 3.4.6
linked bar chart	See paragraph 3.4.10
local regulations	Laws of authorities in the locality of the works
lockout	Exclusion of the workforce from the works by the employer

logic	See paragraph 3.8.26
logic tracing	The process of tracking a path by reference to driving relationships (q.v.)
logical interface	A predecessor or successor
logistics	Management of the flow of resources from procurement to completion of the works
long lag	See paragraph 3.8.69
longest path	See 'critical path'
machines	See 'plant'
mandatory-project-finish	See paragraph 3.8.48.1
manually applied constraint	A constraint applied to a network arising other than as a result of the logic (q.v.); see also paragraph 3.8.67
milestone	See paragraph 4.6.8
milestone monitoring	See paragraph 4.6.8
milestone schedule	See paragraph 4.6.8
mitigation	Action taken to alleviate predictable loss, expense or delay
mobilisation period	The lapsed time needed to assemble resources between instructions issued to perform a task and the commencement of a task
moderate constraints	See paragraph 3.8.47
Monte Carlo analysis	See paragraphs 3.8.55.3 to 3.8.55.9
must-finish-on	See paragraph 3.8.48.1
must-start-on	See paragraph 3.8.48.1
named subcontractor	A subcontractor selected from a list provided by the employer (q.v.) but employed by the contractor (who usually takes responsibility for the subcontractor's performance), usually for work of a specialised nature
negative float	See paragraph 3.8.53
negative lag	See paragraphs 3.8.41 and 3.8.70
network	A schedule in which the activities are linked by their logical predecessors and successors
network diagram	See paragraphs 3.4.8 to 3.4.10.2
node	The junction between arrows in a network representing an event; see also paragraphs 3.4.8 to 3.4.10.2
nominated subcontractor	A subcontractor selected by the employer (q.v.), (who usually takes responsibility for the subcontractor's performance) but employed by the contractor, usually for work of a specialised nature
non-contiguous activity	See 'interruptible activity'
non-driving relationship	Logic required to close a network and ensure the absence of open ends (q.v.) and which is not on critical path (q.v.)

occupational commissioning	Rendering the works fit for use by the end user
occupational-commissioning schedule	See paragraph 3.3.6
open end	See paragraph 3.8.68
organisation-breakdown structure	The relationships which define the responsibilities of administrative and other parties for the execution of a task; see also paragraph 3.8.1.7
organising	Putting into the required order
out-turn cost	The ultimate cost of a project: the tender cost plus the cost of variations and compensation for loss and/or expense, including consultants' fees, planning fees and licenses etc.; see also paragraph 2.1.5
overload	The status of a project in which, in order to meet a duration, resources are required in excess of that planned
overtime	Time required to be worked in excess of the regular or normal hours of work
owner	See 'employer'
pacing	Slowing down of work for the reduction of resources to keep pace with delayed work
partial possession	The use by the employer (q.v.) of a part of the works before project completion
planning	See 'project planning'
planning method statement	See paragraph 2.6
planning strategy	See paragraph 2.1
plant	Mechanical equipment
possession	Control of an area (q.v.) or zone (q.v.) of operation
precedence diagram	A network in which the activity is on the node
predecessor	That part of an activity which must start or finish before another activity logically can start or finish
preferential logic	See paragraph 3.8.28
prime cost sum	A contingent sum of money included in the contract sum for work to be carried out by a specialist subcontractor or for materials yet to be specified
production records	See 'progress records'
productivity quotients	The rate at which work can be accomplished with a given resource for a given activity
programme	The time-control document required by some forms of contract, usually in printed form; see also 'schedule'
progress monitoring	See paragraph 4.6

progress records	Records of progress achieved; see also paragraphs 2.7, 4.3.5 and 4.3.6
progress update	See paragraph 2.4
project control	See paragraph 1.8
project manager	The party employed by the employer (q.v.) to coordinate the input of the other consultants and contractors, in connection with the project
project planning	See paragraph 1.6
project scheduler	See paragraph 1.7
project scope	All the work to be carried out to reach the project's objectives: there could well be several contracts involved in project delivery from project viability, feasibility, design and implementation/construction
provisional sum	A contingent sum of money included in the contract sum for work of which the detail cannot be fully described at the time of tender
rebar	See 'reinforcement'
record keeping	See paragraph 2.7
record retrieval	Access for use of stored information; see also paragraph 4.3.1.1
recovery	Making up of lost time at the contractor's expense; see also 'acceleration'
reinforcement	Steel bars bedded in concrete to add tensile strength; also known as 'rebar'
remaining time	The duration planned to elapse before an activity is completed
repetitive cycle	A sequence which is carried out more than twice
resource	Anything necessary for the achievement of work but typically, materials, labour, plant, space, cost
resource-allocation data	Information about the use of a resource
resource levelling	See paragraphs 3.5.2.10 and 3.8.15.7
resource logic	See paragraph 3.8.29
resource planning	See paragraph 3.5
resource scheduling	See paragraph 3.5
resource smoothing	See 'resource levelling'
rework	Repair of defective work
risk manager	The title given to the party responsible for managing the risk register
risk register	Schedule of foreseeable risks, likelihood of occurrence, possible consequences and planned remedial action
schedule	The time-model for the work
schedule density	See paragraph 3.7.11
schedule design	See paragraph 3.7

schedule integrity	The character necessary for the schedule to perform dynamically so as to properly calculate the consequences of change; see also paragraph 3.8.66
schedule review	See paragraph 4.2
scheduler	See paragraph 1.7
scheduling at high density	See paragraph 3.7.14
scheduling at low density	See paragraph 3.7.12
scheduling at medium density	See paragraph 3.7.13
scheduling options	Software switches which select different computational algorithms (q.v.); see also paragraph 3.8.72
SCL Protocol	The Delay and Disruption Protocol, published by the Society of Construction Law, London, in 2002
sectional completion date	The date by which a defined part of the works is contractually bound to be completed
separate contractor	A contractor employed to carry out work under a different contract from that for the main works
short-term, look-ahead report	A narrative describing the high-density part of the schedule from the data date (q.v.) going forward
shuttering	See 'formwork'
simple projects	See paragraph 1.5.2
smoothing	See 'resource levelling'
sorting	The process of defining the order in which values will be displayed in a database report
spreadsheet	An electronic grid of cells in which alphanumeric data or formulae can be located and interrelated for listing, filtering, sorting, organisation and calculation
standard outputs	Published productivity quotients (q.v.); see also paragraph 3.5
start-no-earlier-than	See paragraph 3.8.47.1
start-no-later-than	See paragraph 3.8.47.1
start-to-finish	See paragraph 3.8.35
start-to-start	See paragraph 3.8.32
statutory approvals	Permissions which are required by law
statutory undertaker	A company authorised by law to provide services usually concerned with transport or the provision of water, gas, electricity and so on. See also 'utilities'
stretched activity	See 'interruptible activity'
strike	A group action withdrawing labour from the work
strike formwork	Remove formwork (q.v.)

submittal	An application for approval or consent
sub-project	A part of the works with a start and completion date
successor	That part of an activity which logically cannot start or finish until after another activity has started or finished
target schedule	A schedule against which a departure from it is measured
temporary works	Work which must be carried out in order to construct the permanent works, but which is not intended to remain
tender	An offer to carry out work for compensation under a contract; also known as a 'bid'
tender schedule	See paragraph 3.3.4
testing and commissioning	The process of validating and adjusting the permanent works, or any part of it, and rendering it fit for use
third-party issues	Those matters requiring action relating to work to be carried out by other contractors or companies in connection with the works
third-party projects	Work carried out other than that which is the subject of the contract between the contractor and employer (q.v.)
time-chainage diagram	See paragraph 3.4.7
time contingency	A period of time allowed for work or suspension of work which may or may not be required
time-contingency buffer	See 'time contingency'; usually used in connection with critical chain management
time-impact analysis	A method of calculating the likely effect of an intervening event, taking into account progress achieved prior to the occurrence of the event
time-management strategy	See Part 2
time-model	See Part 3
time now	See 'data date'
topping-out	The term given to the completion of the last part of a structural enclosure
total float	See paragraph 3.8.52
trade	A particular specialised type of work
trade-package contractors	Contractors engaged to carry out a defined part of the works under a construction-management contract
trailing open end	An activity without a successor to its finish; see also paragraph 3.8.68
triangular distribution	See paragraphs 3.8.55.5 to 3.8.55.7
turnkey	See 'engineer, procure and construct'
unexpired contingencies	A time allowance which is unused; see also 'contingency'
utilities	The public-company suppliers of water, gas, electricity, communications and other publicly available services. See also 'statutory undertaker'

variation	An instructed change as defined under the contract
windows	The expression used to describe a time analysis related to predetermined periods of time; see also paragraph 4.5.15.6
work-breakdown structure	See paragraph 3.7.8
work pattern	The name given to the sequence of working periods and non-working periods in a working day
work-type definition	A description of work in the detail which renders it unique amongst other work descriptions of like type
working schedule	See paragraph 3.3.5
zero-free-float	See paragraph 3.8.47.1
zero-total-float	See paragraph 3.8.48.1
zonal logic	See paragraph 3.8.30
zone of operation	A division of the work for the purposes of management and control

Index

In this index figures are indicated in **bold** type. Appendices are indicated by a.